ELECTRICIAN'S GUIDE TO EMERGENCY LIGHTING

Published by The Institution of Engineering and Technology, London, United Kingdom

The Institution of Engineering and Technology is registered as a Charity in England & Wales (no. 211014) and Scotland (no. SCO38698).

The Institution of Engineering and Technology is the new institution formed by the joining together of two great institutions; the IEE (Institution of Electrical Engineers) and the IIE (The Institution of Incorporated Engineers). The new Institution is the inheritor of the IEE brand and all its products and services, such as this one, which we hope you find useful.

Copies of this publication may be obtained from:
The Institution of Engineering and Technology
PO Box 96, Stevenage, SG1 2SD, UK
Tel: +44 (0)1438 767328
Email: sales@theiet.org
www.theiet.org/wiringbooks

While the author, publisher and contributors believe that the information and guidance given in this work are correct, all parties must rely upon their own skill and judgement when making use of them. The author, publisher and contributors do not assume any liability to anyone for any loss or damage caused by any error or omission in the work, whether such an error or omission is the result of negligence or any other cause. Where reference is made to legislation it is not to be considered as legal advice. Any and all such liability is disclaimed.

Extracts from BS 5266-1 (2011) are reproduced within this Publication with the permission of the BSI Standards Limited. Copyright subsists in all BSI publications. British Standards can be obtained in PDF or hard copy formats from the BSI online shop: www.bsigroup.com/shop or by contacting BSI Customer Services for hard copies only: Tel: +44 (0)20 8996 9001, Email: cservices@bsigroup.com.

ISBN 978-1-84919-771-7
eISBN 978-1-84919-772-4

Typeset in the UK by Carnegie Book Production, Lancaster
Printed in the UK by Polestar Wheaton

Contents

Cooperating organisations

The IET acknowledges the invaluable contribution made by the following organisations in the preparation of this Guide.

BEAMA Installations Ltd
Eur Ing M.H. Mullins BA CEng FIET
P. Sayer IEng MIIE GCGI

British Standards Institution

Department for Communities and Local Government
A. Burd

Electrical Contractors' Association
G. Digilio IEng FIET ACIBSE MSLL

Electrical Contractors' Association of Scotland t/a SELECT
D. Millar IEng MIIE MILE

Electrical Safety Council

Fire and Security Association
M. Turner

Health and Safety Executive
K. Morton CEng MIET

Institution of Engineering and Technology
M. Coles BEng(Hons) MIEE
P.E. Donnachie BSc CEng FIET

Scottish Government Building Standards Division
C. Donnelly

Author
P.R.L. Cook CEng FIET

The author would like to record special thanks to Eur Ing Leon Markwell MSc BSc CEng MIET MCIBSE LCGI of the ECA for his particular assistance in preparing the publication.

Acknowledgements

References to British Standards are made with the kind permission of BSI. Complete copies can be obtained by post from:

BSI Customer Services
389 Chiswick High Road
London W4 4AL

For all enquiries contact:

Tel: +44 (0)20 8996 9001
Fax: +44 (0)20 8996 7001

Email: cservices@bsi-global.com
www.bsi-global.com/en/Shop/

References to Building Regulations, Approved Documents and guidance are made with the kind permission of the Department for Communities and Local Government. Downloads of Approved Documents are available from the Planning Portal: www.planningportal.gov.uk

Preface

The *Electrician's Guide to Emergency Lighting* is one of a number of publications prepared by the IET to provide guidance on electrical installations in buildings. This publication is concerned with emergency lighting and in particular emergency escape lighting and must be read in conjunction with the legislation, Approved Document B and the relevant British Standards, in particular BS 5266.

Designers and installers should always consult these documents to satisfy themselves of compliance.

It is expected that persons carrying out work in accordance with this Guide will be competent to do so, competence being a statutory requirement of the Electricity at Work Regulations 1989 for those engaged in electrical work.

Legislation 1

1.1 Introduction

It is necessary to start with a few basic definitions of emergency lighting terms; see diagram below:

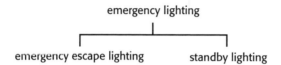

Emergency escape lighting is divided into three subcategories:

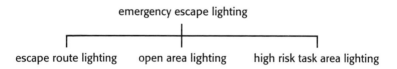

These terms are described more fully in section 3.3.

Common law and much legislation impose duties of care upon those responsible for buildings, in particular employers. If precautions and provisions are such that no person is harmed no problem arises. This must be the prime objective, to determine what is necessary for the safety of persons using the facility and to implement the requirements.

Laws, regulations, standards and guidance documents translate this general duty into specific requirements: numbers, locations, fire performance etc. However, these recommendations must not be allowed to override what the designer, installer, operator with their unique knowledge of the facility know to be necessary.

There is much legislation that is applicable to emergency lighting; general safety legislation includes:

i   The Health and Safety at Work etc. Act 1974
ii The Management of Health and Safety at Work Regulations 1999 (Statutory Instrument 1999 No. 3242)

iii The Workplace (Health, Safety and Welfare) Regulations 1992 (Statutory Instrument 1992 No. 3004)

iv The Health and Safety (Safety Signs and Signals) Regulations 1996 (Statutory Instrument 1996 No. 341).

Legislation with specific requirements for emergency lighting includes:

v The Building Regulations 2010 as amended (England and Wales) (Statutory Instrument 2010 No. 2214)

vi The Regulatory Reform (Fire Safety) Order 2005 (Statutory Instrument 2005 No. 1541)

vii The Cinematograph (Safety) Regulations 1955 (Statutory Instrument 1955 No. 1129) as amended.

These are discussed below.

1.2 The Health and Safety at Work etc. Act 1974

Together with the common law general duty of care of everyone for his neighbour, the most fundamental legislation with respect to emergency lighting is the Health and Safety at Work etc. Act 1974. It includes comprehensive requirements for health, safety and welfare at work, the control of dangerous substances and certain emissions into the atmosphere. The Act empowers the Secretary of State to make regulations, generally referred to as health and safety regulations, and issue approved codes of practice. The Secretary of State has made much use of this power. In many respects, the legislation concerned with emergency lighting is similar to that for fire safety.

The general duties are given in section 1:

1. Preliminary

(1) The provisions of this Part shall have effect with a view to –

(a) securing the health, safety and welfare of persons at work;

(b) protecting persons other than persons at work against risks to health or safety arising out of or in connection with the activities of persons at work;

The status of approved codes of practice is given in section 17 as follows:

17. Use of approved codes of practice in criminal proceedings

(1) A failure on the part of any person to observe any provision of an approved code of practice shall not of itself render him liable to any civil or criminal proceedings; but where in any criminal proceedings a party is alleged to have committed an offence by reason of a contravention of any requirement or prohibition imposed by or under any such provision as is mentioned in section 16(1) being a provision for which there was an approved code of practice at the time of the alleged contravention, the following subsection shall have effect with respect to that code in relation to those proceedings.

1.3 The Management of Health and Safety at Work Regulations 1999 (SI 1999 No. 3242)

The Management of Health and Safety at Work Regulations include a requirement for risk assessment.

3. Risk assessment

(1) Every employer shall make a suitable and sufficient assessment of –

(a) the risks to the health and safety of his employees to which they are exposed whilst they are at work; and

(b) the risks to the health and safety of persons not in his employment arising out of or in connection with the conduct by him of his undertaking,

for the purpose of identifying the measures he needs to take to comply with the requirements and prohibitions imposed upon him by or under the relevant statutory provisions.

This requirement has particular relevance to fire systems, including emergency lighting.

European Council directives require emergency lighting, and the consequence is that many of the health and safety regulations issued by the Secretary of State with the authority of the Health and Safety at Work Act are also implementing European Council directives as well as meeting UK initiatives; see chapter 2, Building Regulations.

1.4 The Workplace (Health, Safety and Welfare) Regulations 1992 (SI 1992 No. 3004)

In regulation 8 of the Workplace (Health, Safety and Welfare) Regulations there is a requirement that suitable and sufficient emergency lighting shall be provided in any room in circumstances in which persons at work are specially exposed to danger in the event of failure of artificial lighting. This is a general requirement to provide standby lighting. See regulation 8 below.

8. Lighting

(1) Every workplace shall have suitable and sufficient lighting.

(2) The lighting mentioned in paragraph (1) shall, so far as is reasonably practicable, be by natural light.

(3) Without prejudice to the generality of paragraph (1), suitable and sufficient emergency lighting shall be provided in any room in circumstances in which persons at work are specially exposed to danger in the event of failure of artificial lighting.

1.5 The Health and Safety (Safety Signs and Signals) Regulations 1996 (SI 1996 No. 341)

Where the risk assessment required by the Management of Health and Safety at Work Regulations indicates a need for any safety signs or signals, the Health and Safety (Safety Signs and Signals) Regulations require that suitable signs be installed with, if necessary, a standby electricity supply.

Regulation 4 and relevant paragraphs of schedule 1 of the Health and Safety (Safety Signs and Signals) Regulations 1996 are reproduced below.

4. Provision and maintenance of safety signs

(1) Paragraph (4) shall apply if the risk assessment made under paragraph (1) of regulation 3 of the Management of Health and Safety at Work Regulations 1992 indicates that the employer concerned, having adopted all appropriate techniques for collective protection, and measures, methods or procedures used in the organisation of work, cannot avoid or adequately reduce risks to employees except by the provision of appropriate safety signs to warn or instruct, or both, of the nature of those risks and the measures to be taken to protect against them.

(4) Where this paragraph applies, the employer shall (without prejudice to the requirements as to the signs contained in regulation 11(2) of the Offshore Installations (Prevention of Fire and Explosion, and Emergency Response) Regulations 1995) –

  (a) in accordance with the requirements set out in Parts I to VII of Schedule 1, provide and maintain any appropriate safety sign (other than a hand signal or verbal communication) described in those Parts, or ensure such sign is in place; and

  (b) subject to paragraph (5), in accordance with the requirements of Parts I, VIII and IX of Schedule 1, ensure, so far as is reasonably practicable, that any appropriate hand signal or verbal communication described in those Parts is used; and

  (c) provide and maintain any safety sign provided in pursuance of paragraph (6) or ensure such sign is in place.

SCHEDULE 1 – Regulation 4(4) and (5)

PART I

MINIMUM REQUIREMENTS CONCERNING SAFETY SIGNS AND SIGNALS AT WORK

8. Signs requiring some form of power must be provided with a guaranteed emergency supply in the event of a power cut, unless the hazard has thereby been eliminated.

9. The triggering of an illuminated sign and/or acoustic signal indicates when the required action should start; the sign or signal must be activated for as long as the action requires. Illuminated signs and acoustic signals must be reactivated immediately after use.

10. Illuminated signs and acoustic signals must be checked to ensure that they function correctly and that they are effective before they are put into service and subsequently at sufficiently frequent intervals.

The Health and Safety Executive has issued guidance on these regulations in publication L64. This guidance details the recommended signs and signals, including emergency escape and fire-fighting signs. Chapter 7 reproduces many of the recommended signs.

1.6 The Building Regulations 2010 (SI 2010 No. 2214)

Requirement B1 'Means of warning and escape' of schedule 1 of the Building Regulations is reproduced in chapter 2.

The Secretary of State issues 'Approved Documents' that offer practical guidance on the requirements of the Building Regulations. There is no obligation to adopt any particular solution in an Approved Document, if the designer wishes to meet the requirement in another way. However, all the most common situations are dealt with in a practical way. Approved Document B provides guidance on the requirements of schedule 1 and regulation 7 of the Building Regulations and is discussed in chapter 2.

1.7 The Regulatory Reform (Fire Safety) Order 2005 (SI 2005 No. 1541)

The Regulatory Reform (Fire Safety) Order 2005 is intended to rationalize and simplify much of the legislation concerning fire safety at work in England and Wales; it amends or replaces over 100 pieces of legislation including many local authority acts.

For new buildings and substantial alterations, the Building Regulations impose requirements for emergency lighting associated with fire safety, see chapter 2.

The Regulatory Reform (Fire Safety) Order 2005 applies to the majority of premises and workplaces in the United Kingdom, but generally does not apply to dwellings.

The Order requires the responsible person to carry out a fire risk assessment and prepare a policy document for fire safety including escape lighting. Escape procedures must be developed, staff trained, means of escape prepared including escape signs, escape notices, emergency lighting, fire detection and alarm systems installed and fire-fighting equipment positioned as necessary.

Regulation 14 has specific requirements for emergency routes and exits and is reproduced below.

14. Emergency routes and exits

(1) Where necessary in order to safeguard the safety of relevant persons, the responsible person must ensure that routes to emergency exits from premises and the exits themselves are kept clear at all times.

(2) The following requirements must be complied with in respect of premises where necessary (whether due to the features of the premises, the activity carried on there, any hazard present or any other relevant circumstances) in order to safeguard the safety of relevant persons –

(a)–(f) *omitted*
(g) emergency routes and exits must be indicated by signs; and
(h) emergency routes and exits requiring illumination must be provided with emergency lighting of adequate intensity in the case of failure of their normal lighting.

The Order revokes much legislation, see table below:

Instrument	Reference	Extent of revocation
The Fire Certificate (Special Premises) Regulations 1976	SI 1976 No. 2003	The whole Regulations
The Fire Precautions (Workplace) Regulations 1997	SI 1997 No. 1840	The whole Regulations
The Fire Precautions (Workplace) (Amendment) Regulations 1999	SI 1999 No. 1877	The whole Regulations
The Management of Health and Safety at Work Regulations 1999	SI 1999 No. 3242	In regulations 1(2), 3(1), 7(1), 11(1)(a), 11(1)(b), 12(1)(b) the words 'and by Part II of the Fire Precautions (Workplace) Regulations 1997' in each place where they occur. In regulation 10(1)(c) the words from 'and the measures' to 'Regulations 1997'. In regulation 10(1)(d) the words 'and regulation' to 'Regulations 1997'. In regulations 11(2) and 12(2) the words in brackets. Regulation 28.

1.8 The Cinematograph (Safety) Regulations 1955 (SI 1955 No. 1129) as amended

The regulations require 'Safety lighting'. Regulation 16. -(1) states:

Safety lighting

16. -(1) In addition to the general lighting, means of illumination adequate to enable the public to see their way out of the premises without assistance from the general lighting shall be provided –

(a) in the auditorium and all other parts of the building to which the public are admitted;

(b) in all passages, courts, ramps and stairways to which the public have access and which lead from the auditorium to outside the premises;

(c) for the illumination of all notices indicating exits from any part of the premises to which the public are admitted.

(2) The safety lighting shall be kept on at all times when the public are on the premises except in those parts of the premises which are lit equally well by daylight.

The requirements of more recent legislation would seem to include rather more comprehensive requirements to protect all people using the premises. Cinemas are included within the scope of BS 5266-1 (see clause 9.3.4).

Building Regulations

2

2.1 Introduction

2.1.1 The Building Regulations 2010 (England and Wales)

The Building Regulations 2010 (England and Wales), SI 2010 No. 2214, apply to building work as described in regulation 3 of the Building Regulations. In the regulations, building work means:

a the erection or extension of a building;

b the provision or extension of a controlled service or fitting in or in connection with a building;

c the material alteration of a building or a controlled service or fitting;

d work relating to a material change of use;

e insulation filling of cavity walls;

f work involving the underpinning of a building.

The fire safety requirements of the Building Regulations are given in part B of schedule 1 to the regulations. The particular requirement with respect to emergency lighting is requirement B1 'Means of warning and escape'.

▼ **Figure 2.1** Requirement B1 of schedule 1 to the Building Regulations

Requirement	Limits on application
Means of warning and escape **B1,** The building shall be designed and constructed so that there are appropriate provisions for the early warning of fire, and appropriate means of escape in case of fire from the building to a place of safety outside the building capable of being safely and effectively used at all material times.	Requirement B1 does not apply to any prison provided under section 33 of the Prisons Act 1952 (power to provide prisons etc.).

Guidance on schedule B1 to the Building Regulations is given in Approved Document B: Fire safety. This Approved Document comes in two volumes:

> Volume 1 – dwellinghouses
> Volume 2 – buildings other than dwellinghouses.

The guidance given in this chapter is based on the guidance in Volumes 1 and 2 of the Approved Document.

2.1.2 Wales

On 31 December 2011 the power to make building regulations for Wales was transferred to Welsh Ministers. This means Welsh Ministers will make any new building regulations or publish any new building regulations guidance applicable to Wales from that date.

The Building Regulations 2010 and related guidance, including Approved Documents as at that date, will continue to apply in Wales until Welsh Ministers make changes to them. For example, the 2006 version of Approved Document P applies at the date of publication of this Guide.

2.1.3 Status of Approved Documents

The 'Planning Portal' advises: The Approved Documents are intended to provide guidance for some of the more common building situations. However, there may well be alternative ways of achieving compliance with the requirements. Thus there is no obligation to adopt any particular solution contained in an Approved Document if it is preferred to meet the relevant requirement in some other way.

They are given legal status by the Building Act 1984. Regulation 6 states:

6. Approval of documents for purposes of building regulations

(1) For the purpose of providing practical guidance with respect to the requirements of any provision of building regulations, the Secretary of State or a body designated by him for the purposes of this section may –

(a) approve and issue any document (whether or not prepared by him or by the body concerned), or

(b) approve any document issued or proposed to be issued otherwise than by him or by the body concerned,

if in the opinion of the Secretary of State or, as the case may be, the body concerned the document is suitable for that purpose.

Regulation 7 states:

7. Compliance or non-compliance with approved documents

(1) A failure on the part of a person to comply with an approved document does not of itself render him liable to any civil or criminal proceedings; but if, in any proceedings whether civil or criminal, it is alleged that a person has at any time contravened a provision of building regulations –

  (a) a failure to comply with a document that at that time was approved for the purposes of that provision may be relied upon as tending to establish liability, and

  (b) proof of compliance with such a document may be relied on as tending to negative liability.

(2) In any proceedings, whether civil or criminal –

  (a) a document purporting to be a notice issued as mentioned in section 6(3) above shall be taken to be such a notice unless the contrary is proved, and

  (b) a document that appears to the court to be the approved document to which such a notice refers shall be taken to be that approved document unless the contrary is proved.

2.2 Approved Document B, Volume 1 – dwellinghouses

Dwellinghouse: A unit of residential accommodation occupied (whether or not as a sole or main residence):

  a by a single person or by people living together as a family
  b by not more than six residents living together as a single household, including a household where care is provided for residents.

Dwellinghouse does not include a flat or a building containing a flat.

As the scope of Volume 1 is dwellinghouses, and it does not include flats, the only reference to escape lighting is a definition.

2.3 Approved Document B, Volume 2 – buildings other than dwellinghouses

2.3.1 Lighting of escape routes

(para 5.36 of Vol. 2)

Volume 2 of Approved Document B recommends that all escape routes should have adequate artificial lighting, and that certain escape routes and areas should also have escape lighting to illuminate the escape route should the mains supply fail, see Table 2.1 overleaf. The lighting to escape stairs should be on a separate circuit from the circuits supplying the lighting to other parts of the escape route. The intention is to provide for a more reliable supply to escape stairs and designers should bear this intention in mind when designing the lighting.

▼ **Table 2.1**   Provisions for escape lighting (Table 9 of App Doc B, Vol. 2)

Purpose group of the building or part of the building	Areas requiring escape lighting
1. Residential	All common escape routes*, except in 2-storey flats
2. Office, storage and other non-residential	a. Underground or windowless accommodation b. Stairways in a central core or serving storey(s) more than 18 m above ground level c. Internal corridors more than 30 m long d. Open-plan areas of more than 60 m^2
3. Shop, commercial and car parks	a. Underground or windowless accommodation b. Stairways in a central core or serving storey(s) more than 18 m above ground level c. Internal corridors more than 30 m long d. Open-plan areas of more than 60 m^2 e. All escape routes* to which the public are admitted (except in shops of three or fewer storeys with no sales floor more than 280 m^2 provided that the shop is not a restaurant or bar)
4. Assembly and recreation	All escape routes*, and accommodation except for: a. accommodation open on one side to view sport or entertainment during normal daylight hours
5. Any purpose group	a. All toilet accommodation with a floor area over 8 m^2 b. Electricity and generator rooms c. Switchroom/battery room for emergency lighting system d. Emergency control room

*   Including external escape routes.

2.3.2 Exit signs

(para 5.37 of Vol. 2)

Volume 2 of Approved Document B recommends that, except within a flat, an exit sign should distinctively and conspicuously mark every escape route. The sign should be of an adequate size complying with the Health and Safety (Safety Signs and Signals) Regulations 1996. Escape routes in ordinary use, that is typically the main entrance door of a building, are excluded from this recommendation.

There is a recommendation that, in general, signs should conform to BS 5499-1 *Graphical symbols and signs. Safety signs, including fire safety signs. Specification for geometric shape, colours and layout* (now replaced by BS ISO 3864-1:2011).

Chapter 7 of this publication provides information on the Safety Signs and Signals Regulations and on the British Standard.

2.3.3 Critical electrical circuits

(para 5.38 of Vol. 2)
(clause 8.2.2 of BS 5266-1)

Where it is critical for electrical circuits to be able to continue to function during a fire, such as circuits to luminaires and signs from a central standby supply, Approved Document B calls for 'protected circuits'. A protected circuit for operation of equipment in the event of fire is required by Approved Document B to:

a consist of cable meeting the requirements for PH 30 classification when tested to BS EN 50200:2006, Annex E, or an equivalent standard,

b follow a route selected to pass only through parts of the building in which the fire risk is negligible, and

c be separate (i.e. segregated) from any circuit provided for another purpose.

(**Note:** PH is a measure of a cable's ability to continue to perform at a test notional temperature of 842 °C for:

15 minutes	PH 15
30 minutes	PH 30
60 minutes	PH 60
90 minutes	PH 90
120 minutes	PH120

The tests are described in BS EN 50200. The water spray requirements are stated in Annex E of BS EN 50200 for standard emergency lighting cables and in BS 8434-2 for enhanced emergency lighting systems.)

The Building Regulations add that in large or complex buildings where the fire protection system needs to operate for extended periods, guidance should be sought from BS 5266-1 (and BS 5839-1 and BS 7346-6).

BS 5266-1 describes two levels of fire resistance:

i standard performance cables and cable systems with an inherently high resistance to fire, and

ii enhanced performance cables and cable systems with an inherently high resistance to fire.

Standard performance cables and cable systems (with high resistance to fire) are ordinarily used and high performance cables and cable systems (with inherently high resistance to fire) are used, for example, when evacuation is staged and or the building is unsprinklered, please see section 5.3.

Standard cables and cable systems are required generally to have a duration of survival of 60 minutes and enhanced cables or cable systems a duration of survival of 120 minutes. Section 5.3 describes cables and cable systems intended to meet these duration of survival times.

Emergency lighting standards

3

3.1 European standards

The European standards for emergency lighting are:

- ▶ BS EN 1838:2013 Lighting applications – Emergency lighting
- ▶ BS EN 50172:2004, also numbered BS 5266-8:2004 Emergency escape lighting systems.

The United Kingdom, as a member of the European standards body CENELEC, is obliged to give these standards the status of a national standard without any alterations.

British Standard 5266-1 *Emergency lighting – Part 1: Code of practice for the emergency lighting of premises*, was revised (2005) to remove any requirements that are within the scope of the two European standards.

3.2 British Standards

The British Standards Institution standards series for emergency lighting is BS 5266:

- ▶ BS 5266-1:2011 *Emergency lighting. Code of practice for the emergency escape lighting of premises*
- ▶ BS 5266-2:1998 *Emergency lighting. Code of practice for electrical low mounted way guidance systems for emergency use*
- ▶ BS 5266-4:1999 *Emergency lighting. Code of practice for design, installation, maintenance and use of optical fibre systems*
- ▶ BS 5266-5:1999 *Emergency lighting. Specification for component parts of optical fibre systems*
- ▶ BS 5266-6:1999 *Emergency lighting. Code of practice for non-electrical low mounted way guidance systems for emergency use. Photoluminescent systems*
- ▶ BS EN 1838:2013 *Lighting applications – Emergency lighting*
- ▶ BS 5266-8:2004, BS EN 50172:2004 *Emergency escape lighting systems.*

Product (equipment) standards include:

- ► BS EN 60598-2-22:1998 +A2 Luminaires for emergency lighting
- ► BS EN 50171:2001 Central power supply systems
- ► BS EN 62034:2012 Automatic test system for battery powered emergency escape lighting.

The three core standards, parts 1 and 8 of BS 5266 and BS EN 1838, need to be considered together, see Figure 3.1.

Part 1 is a code of practice making recommendations; parts 7 and 8 are the European system standards. The system standards are supported by product standards, particularly those listed above and shown in Figure 3.1.

Note: Persons carrying out emergency lighting designs will need to obtain a copy of parts 1, 7 and 8 as the requirements of the system standards (parts 7 and 8) are not reproduced in the code of practice (part 1).

▼ **Figure 3.1**   Relationship between the core emergency lighting standards (parts 1, 7 and 8 of BS 5266) and product standards

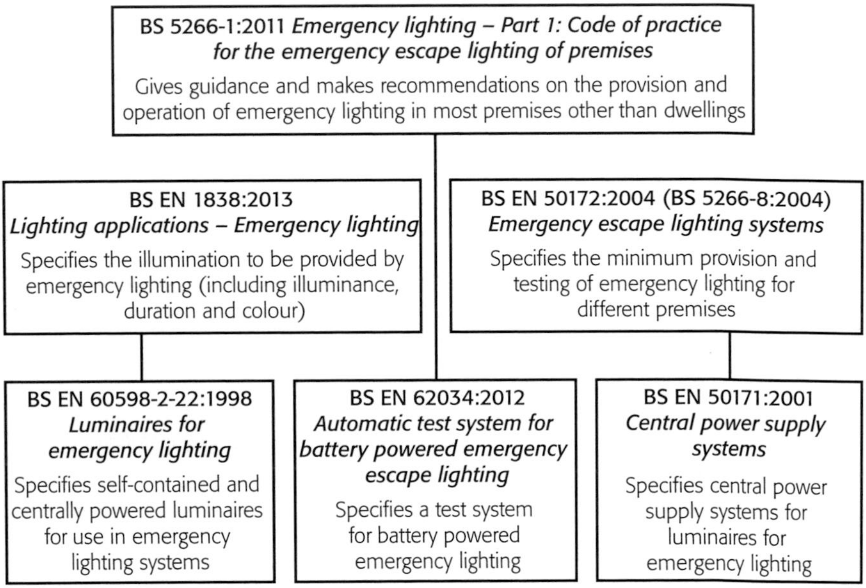

BS 5266-1:2011 *Emergency lighting – Part 1: Code of practice for the emergency escape lighting of premises*
Gives guidance and makes recommendations on the provision and operation of emergency lighting in most premises other than dwellings

BS EN 1838:2013
Lighting applications – Emergency lighting
Specifies the illumination to be provided by emergency lighting (including illuminance, duration and colour)

BS EN 50172:2004 (BS 5266-8:2004)
Emergency escape lighting systems
Specifies the minimum provision and testing of emergency lighting for different premises

BS EN 60598-2-22:1998
Luminaires for emergency lighting
Specifies self-contained and centrally powered luminaires for use in emergency lighting systems

BS EN 62034:2012
Automatic test system for battery powered emergency escape lighting
Specifies a test system for battery powered emergency lighting

BS EN 50171:2001
Central power supply systems
Specifies central power supply systems for luminaires for emergency lighting

3.3 Definitions

The definitions in the standards are not completely identical. The definitions below are mostly those of BS EN 1838 and BS 5266-8:

Emergency lighting

Lighting provided for use when the supply to the normal lighting fails (clause 3.4 of BS 5266-1).

Emergency lighting includes emergency escape lighting and standby lighting:

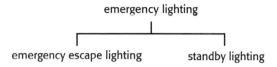

Standby lighting

That part of emergency lighting provided to enable normal activities to continue substantially unchanged (clause 3.7 of BS EN 1838).

Emergency escape lighting

That part of emergency lighting that provides illumination for the safety of people leaving a location or attempting to terminate a potentially dangerous process before doing so (clause 3.3 of BS EN 1838).

Emergency escape lighting includes escape route lighting, open area lighting and high risk task area lighting.

Emergency escape route lighting

That part of emergency escape lighting provided to ensure that the means of escape can be effectively identified and safely used at all times when the location is occupied (clause 3.3 of BS 5266-8; this is called escape route lighting in clause 3.4 of BS EN 1838).

The objective of escape route lighting is to enable safe exit for occupants by providing appropriate visual conditions and direction finding on escape routes and to ensure that fire-fighting and safety equipment can be readily located and used.

Open area (anti-panic)

Areas of undefined escape routes in halls or premises larger than 60 m^2 floor area or smaller areas if there is additional hazard such as use by a large number of people (clause 3.4 of BS 5266-8 and similar clause 3.5 of BS EN 1838).

High risk task area lighting

That part of emergency escape lighting that provides illumination for the safety of people involved in a potentially dangerous process or situation and to enable proper shut down procedures for the safety of the operator and other occupants of the premises (clause 3.6 of BS EN 1838).

3.4 BS 5266 Emergency lighting

3.4.1 Introduction

In this summary of the requirements of the British Standard, the structure of BS 5266 Part 1 (Code of practice for the emergency escape lighting of premises) has been followed. The requirements of Parts 7 and 8, that is the requirements of the European standards, have been inserted as appropriate.

It should be noted that the scope of application of BS 5266-1:2011 includes theatres, cinemas and a wide range of other premises used for entertainment or recreation (see clause 9.3.4 of the standard).

The standard specifies requirements for 'emergency lighting', which by definition is 'lighting provided for use when the supply to the normal lighting fails'. Emergency lighting includes escape lighting and standby lighting. These are discussed further below.

Note: Emergency escape lighting is required not only on complete failure of the supply to the normal lighting but also on a localized failure if such a failure would present a hazard, e.g. a lighting circuit on a stairway.

3.4.2 Standby lighting   (clause 4.5 of BS EN 1838)

Standby lighting, that part of emergency lighting provided to enable normal activities to continue substantially unchanged, is within the scope of the standard, in particular part 7 (BS EN 1838). However, few requirements are listed. When standby lighting is used as part of the emergency escape lighting, it must comply with the requirements for emergency escape lighting. Where standby lighting is provided at an illumination level lower than the minimum normal lighting level then it is to be used for shut down purposes only.

3.4.3 Emergency escape lighting   (clause 4.1 of BS 5266-8)

Emergency escape lighting is that lighting provided in the event of failure of the electricity supply so as to provide sufficient illumination for the safety of people leaving a location including making safe a potentially dangerous process. (Note: not all routes out of a building will be designated escape routes.)

Escape lighting is not designed to allow normal operations to continue in the building. It is designed to allow making safe and safe exit from the building.

The standard requires that the installation shall ensure that emergency escape lighting fulfils the following functions:

a to illuminate escape route signs;
b to provide illumination onto and along such routes as to allow safe movement towards and through the exits provided to a place of safety, including open area lighting;
c to ensure that fire alarm call points and fire equipment provided along the escape routes can be readily located and used;
d to permit operations concerned with safety measures, including 'high risk task area lighting';
e to provide illumination to and at a place of safety.

Figure 3.2 summarises the functional requirements.

▼ **Figure 3.2**

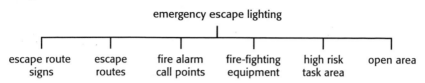

3.4.4 High risk task area lighting

High risk task area emergency lighting is provided for the safety of people involved in potentially high risk work or high risk situations so as to avoid hazards and to allow the proper shut down of work that might otherwise result in danger.

3.4.5 Open area (anti-panic) lighting

Open area lighting, sometimes called anti-panic lighting, is installed to reduce the likelihood of panic in relatively large areas (over 60 m²) with perhaps undefined escape routes such as in halls. It should be considered for smaller areas, if the number of people in the area is likely to cause disorientation or panic.

Design

4

4.1 Introduction

(clause 5 of BS 5266-8)

Emergency escape lighting is a part of the fire safety system (alarms, detectors, sprinklers, gas systems, etc.) and is designed as an integral part of the system. See IET publication *Electrician's Guide to Fire Detection and Alarm Systems.*

The strategy for the evacuation of the building must be decided before designs for the fire detection and alarm system and the emergency lighting installation are started. Designers must know the routes to be taken to ensure that they are appropriately indicated and illuminated. The users (employer) will also need to understand the strategy so that they can ensure that the escape routes are kept clear.

In larger or complex buildings it may be necessary to stage the evacuation and the initiation of sounders or other evacuation signals such as voice indications will need to take account of this.

The selection of escape routes is an integral part of the building design requiring, as well as lighting and signs, other provisions for escape to a place of safety (such as suitable exits, panic bolts, etc).

Emergency escape lighting is required to provide:

i escape route lighting including illumination of:
   a escape route signs;
   b escape routes to allow safe movement towards and through the exits provided to a place of safety, including 'open areas' for anti-panic lighting;
   c fire alarm call points and fire equipment provided along escape routes so that they can be located and used;
ii illumination of high risk task areas (see paragraph 3.4.4) in the event of loss of lighting to allow danger to be avoided and shut down procedures to be implemented as necessary (to facilitate operations concerned with safety measures).

The objective of escape route lighting is to:

i facilitate safe escape for occupants by illumination and direction finding of escape routes as necessary, and
ii enable fire-fighting and safety equipment to be readily (clause 4.1 of BS 5266-8) located and used.

(Escape routes are usually identified during building design but will need to be reviewed from time to time. **They must comply with the Building Regulations and be approved by building control and the fire officer.**)

4.1.1 Risk assessment

(introduction to BS 5266-1)

The Management of Health and Safety at Work Regulations 1999 require every employer to carry out a risk assessment of his premises (see section 1.3). The measures necessary to provide for the safety of staff and other persons using the premises will include the provision of safe means of escape, and are likely to require fire detection and alarm, provision for fire fighting, and include emergency lighting. The requirements of legislation (Chapter 1) and the Building Regulations in particular (Chapter 2) must be considered when relevant. The risk assessment will identify the 'high risk task areas'. See IET publication *Electrician's Guide to Fire Detection and Alarm Systems*.

Risk assessment guidance is given on the HSE site:

www.hse.gov.uk/toolbox/fire.htm

The HSE's publication *Fire Safety – An employer's guide*, details the 'risk assessment' procedure to be used, available as a free download.

4.1.2 Failure of normal supply to part of premises

(clause 5.2 of BS 5266-8)

Emergency escape lighting is required to operate in the event of failure of any part of the normal lighting supply.

Non-maintained and combined non-maintained emergency luminaires are required to operate in the event of failure of a normal lighting final circuit. In all situations the local emergency escape lighting must operate in the event of failure of normal supply to the corresponding local area. (For centrally supplied systems this will require a control system with sensors and contactors for each luminaire or circuit.)

If the normal lighting circuits are interleaved such that in a large room or corridor the luminaires are on different circuits, the risk of a complete loss of lighting is reduced.

4.1.3 Response times

Specifically, BS EN 1838 requires:

a emergency escape route lighting to reach 50% of the required (clause 4.2.6)
    illuminance level within 5 s and full required illuminance within 60 s.

b open area (anti-panic) lighting to reach 50% of the required (clause 4.3.6)
    illuminance within 5 s and full required illuminance within 60 s.

c high risk task area lighting to provide the full required illuminance (clause 4.4.6)
    permanently or within 0.5 s depending upon application.

(If standby generators are used for the emergency lighting, they must comply with these requirements.)

4.1.4 Duration

Minimum duration

(clauses 4.2.5, 4.3.5 and 4.4.5 of BS EN 1838)

The minimum duration for emergency lighting is:

a escape routes – 1 h
b open areas (anti-panic) – 1 h
c high risk task areas – time to make safe and escape.

Requirement

One hour should be sufficient time to evacuate the largest and most complex buildings and reach a place of safety. If it were not, the escape strategy would need to be confirmed as the most effective. However, the duration must be sufficient for orderly escape as per the strategy plus allowance for any disruption to orderly evacuation likely to occur in an emergency.

Allowance should be made for some of the escape routes to be cut off and injured persons given attention.

Following the risk assessment, particularly for larger premises, emergency lighting that will remain in operation after the evacuation of the building has been completed may be necessary to enable searches of the premises to be carried out and to allow return to the building after the emergency.

Further guidance is given in BS 5266-1, clause 9.1. A minimum duration of 3h is required if premises will not be evacuated immediately in the event of a supply failure, such as sleeping accommodation or places of entertainment and places that will be reoccupied again as soon as the supply is restored.

BS 5266-1 states "A minimum duration of 1 h should be used only if the premises will be evacuated immediately on supply failure and not reoccupied until full capacity has been restored to the batteries".

4.1.5 Plans and records

(clauses 5.1 and 6 of BS 5266-8)

Plans showing the layout of the building and all

- escape routes,
- fire alarm call points,
- fire-fighting equipment,
- escape signs,
- structural features that may obstruct the escape route,
- open areas for anti-panic lighting,
- high risk task areas,
- additional emergency lighting, and
- standby lighting

need to be prepared before an emergency lighting installation design can begin.

Upon completion of the installation, as-installed drawings of the emergency escape lighting installation are required to be prepared and a copy retained on the premises. The drawings need to identify all luminaires, and all the main components of the installation. There is a general requirement in BS 5266-8 that the drawings be updated as necessary (following any alteration to the installation) and signed by a competent person.

Plans and designs will need to be approved by the fire officer and building control.

4.1.6 Emergency lighting design procedure   (clause 10 of BS 5266-1)

The following procedure may be followed to determine the emergency lighting requirements:

a Carry out a risk assessment. This will identify the hazards (introduction and para 4.1.1 that will require emergency lighting, including high risk task areas.   of BS 5266-1)

b Refer to the evacuation strategy prepared for the fire detection and alarm system (see IET publication *Electrician's Guide to Fire Detection and Alarm Systems*).

c Position signs and luminaires at primary escape locations   (clause 4.1 of BS EN 1838) with direction signs if necessary, see section 4.2.

d Position luminaires at additional locations.   (clause 6.6 of BS EN 1838)

e Add luminaires as necessary to illuminate the escape   (clause 4.2 of BS EN 1838) routes.

f Add luminaires as necessary to illuminate the open areas. (clause 4.3 of BS EN 1838)

g Illuminate high risk task areas.   (clause 4.4 of BS EN 1838)

h Position safety signs. See Figure 4.1 below.

▼ **Figure 4.1**   Design sequence for emergency lighting luminaires

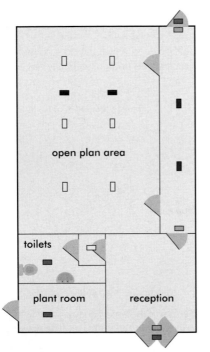

open plan area

toilets

plant room

reception

Design sequence, see 4.1.6
- ▢ c) Position primary emergency escape route signs and luminaires at essential locations (4.2)
- ◼ d) Position additional emergency escape lighting luminaires (4.3)
- ◼ e) Add escape route luminaires as necessary to achieve minimum route illumination levels (4.4)
- ▢ f) Add open area lighting using spacing table (4.5)
- ◼ g) Position high risk task area lighting (4.6)

A decision will have to be made whether centrally supplied luminaires or self-contained luminaires are to be installed. Self-contained luminaires are cheaper to install than centrally supplied, but maintenance costs may be higher; a payback period will need to be given to the client. Smaller installations, say 70 luminaires or fewer, tend to have self-contained luminaires.

4.2 Primary escape route signs and luminaires

(clause 4.1 of BS EN 1838)

4.2.1 Signs

Chapter 7 provides guidance on safety signs as required by the Health and Safety (Safety Signs and Signals) Regulations.

4.2.2 Locations

Generally an escape lighting luminaire complying with BS EN 60598-2-22 is required near (outside) each exit door, and at positions where it is necessary to emphasize potential danger or safety equipment. The positions include the following:

(**Note** to clause 4.1 of BS EN 1838: For the purposes of this clause, 'near' is normally considered to be within 2 m measured horizontally.)

a at each exit door intended to be used in an emergency

b* near stairs so that each flight of stairs receives direct light

# 4

c* near any other change in
level

d* mandatory emergency exits
and safety signs

e* at each change of direction

f at each intersection of
corridors

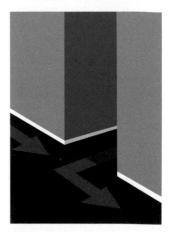

* Note: In locations b, c, d and e, whilst emergency lighting is required a direction sign may not be required.

g (no illustration):   near each final exit and outside the building to a place of safety.

h   near each first-aid post

i   near each piece of fire-fighting equipment and call point

j (no illustration)·   near escape equipment provided for the disabled.

k (no illustration).   near disabled person refuges and disabled person toilet alarm positions.

Where there is no direct sight of an exit and doubt may exist as to its location, a directional sign (or series of signs) shall be provided, placed such that a person moving towards it will be directed towards an emergency exit.

An exit or directional sign shall be in view at all points along the escape route.

All signs marking exits and escape routes shall be uniform in colour and format, and their luminance shall comply with BS 5266-7.

Note: Maintained exit signs should be considered for applications where occupants may be unfamiliar with the building (see 4.8.3).

Escape routes should be unobstructed (see Approved Document M: Access to and use of buildings); where this is not possible obstructions will need to be guarded and illuminated.

4.2.3 Viewing distances (clauses 4.1, 5.4 and 5.5 of BS EN 1838)

Exit signs must be in the direct line of sight of every person in the building and within the viewing distance; see section 4.9 for further information on dimensions etc. If they are not, an illuminated direction sign or signs must be installed to direct persons towards the exit.

4.3 Additional emergency lighting luminaires (clause 6.6 of BS 5266-1)

BS 5266-1 advises that emergency lighting should be provided for:

a External areas in the immediate vicinity of exits to assist dispersal away from the exits to a place of safety, the illumination level being as for escape routes.

b Lift cars. These are not usually used for the escape route, but persons may be trapped in them in the event of a supply failure. Being confined in a small space in the dark without escape is not only unpleasant but may cause harm to those who are nervous or suffer from claustrophobia.

Note: If disabled people are given access to a building, one of their means of escape in emergency conditions may involve use of a lift or a place of safety or stairwell. Emergency lighting as specified for open area (anti-panic) lighting (see section 4.5) is required in lifts in which persons may travel. The emergency lighting can be either self-contained or powered from a central supply, in which case a fire-protected supply will be required.

c Moving stairways and walkways. These should be illuminated as if they were part of an escape route.

d Toilet facilities. Facilities exceeding 8 m^2 gross floor area should be provided with emergency lighting as if they were open areas (see section 4.5).

Toilets for disabled people and any multiple-closet facilities without natural or borrowed* light should have emergency illumination.

Emergency lighting is not mandated in toilets designed for a single able-bodied person or en-suite toilets or bathrooms in hotel bedrooms.

Note:

* Internal windows or clerestories that connect a daylit room to a space inside the core of the building are referred to as borrowed lights and provide a means of getting daylight into places that would not normally be so lit because they have no access to the periphery of the building.

e Motor generator, control, plant and switch rooms. Battery-powered emergency lighting should be provided in all motor generator rooms, control rooms, plant rooms, switch rooms and adjacent to main control equipment associated with the provision of normal and emergency lighting to the premises.

f Covered car parks. The pedestrian escape routes from covered and multi-storey car parks should be provided with emergency lighting.

4.4 Escape route illumination

4.4.1 Illumination levels   (clause 5.1.1 of BS 5266-1, clause 4.2.1 of BS EN 1838)

After positioning primary and additional signs, including supplementary direction signs, the illumination along the escape routes must be such that the means of escape can be identified and safely used in an emergency at any time when the location may be in use.

For routes that are permanently unobstructed and up to 2 m wide, the horizontal illuminance at floor level on the centre line of the escape route should not be less than 1 lx.

First-aid posts and near fire-fighting equipment, if not on the escape route or in an open area, shall be illuminated to 5 lx minimum on the floor.

BS EN 1838 specifies requirements for uniformity of illuminance and the disability glare limits, see below.

Note: For design purposes it may not be reasonable to presume escape routes to be permanently unobstructed.

▼ **Figure 4.2**   Uniformity of illuminance

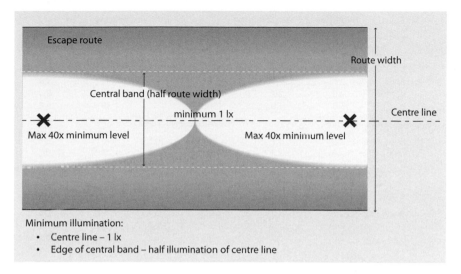

Minimum illumination:
- Centre line – 1 lx
- Edge of central band – half illumination of centre line

For escape routes up to 2 m wide, the horizontal illuminances on the floor along the centre line of an escape route shall be not less than 1 lx and the central band consisting of not less than half of the width of the route shall be illuminated to a minimum of 50% of that value.

Along the centre line of the escape route the ratio of the maximum to the minimum illuminance shall not be greater than 40:1.

These levels of illumination are normally achieved by using the manufacturer's spacing tables to determine the maximum spacing between luminaires, see Table 4.1.

4

▼ **Table 4.1**   Manufacturer's spacing table (typical)

Mode	Mounting height (m)	Lux level directly under	Luminaire spacing (m)			
Escape optic (Asymmetric)			Escape route 2 m wide 1 lux min			
Self-contained	2.5	2.7	–	–	17.1	7.8
	2.8	2.2	–	–	18.6	8.4
	3.0	1.9	–	–	19.8	8.8
Open area optic (Symmetric)			Open (anti-panic) area 0.5 lux min			
Self-contained	2.5	1.7	5.3	10.5	10.5	5.3
	2.8	1.4	5.7	11.5	11.5	5.7
	3.0	1.2	5.8	12.2	12.2	5.8
	4.0	0.67	4.9	12.5	12.5	4.9
Open area optic (Symmetric)			Open area 1 lux min			
Self-contained	2.5	1.7	4.3	9.4	9.4	4.3
	2.8	1.4	3.3	9.3	9.3	3.3
	3.0	1.2	3.2	9.2	9.2	3.2

▼ **Table 4.2**   Disability glare limits (Table 1 of BS EN 1838)

Mounting height above floor level, h (m)	Escape route and open area (anti-panic) lighting maximum luminous intensity, I_{max} (cd)	High risk task area lighting maximum luminous intensity, I_{max} (cd)
$h < 2.5$	500	1000
$2.5 \leq h < 3.0$	900	1800
$3.0 \leq h < 3.5$	1600	3200
$3.5 \leq h < 4.0$	2500	5000
$4.0 \leq h < 4.5$	3500	7000
$h \geq 4.5$	5000	10000

4.4.2 Compartment lighting (clause 5.3 of BS 5266-8)

The illumination by the emergency escape lighting system of a 'compartment' of the escape route, see Figure 4.3, shall be from two or more luminaires so that the failure of one luminaire does not make the directional indication of the system ineffective or result in total darkness in part of the route. For the same reason, two or more luminaires shall be used in each open area with anti-panic lighting.

▼ **Figure 4.3**   Compartment lighting

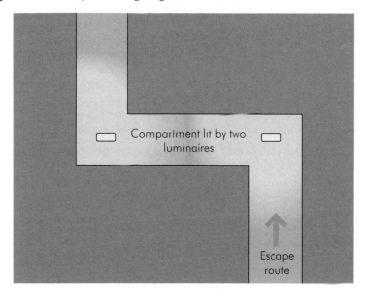

Compartment lit by two luminaires

Escape route

4.5 Open area (anti-panic) lighting (clauses 4.3 of BS EN 1838 and 4.4, 5.3 of BS 5266-8)

The objective of open area (anti-panic) lighting is to reduce the likelihood of panic and to enable safe movement of occupants towards escape routes by providing lighting and direction indication. It is used in open areas where escape routes are ill-defined, such as halls or premises larger than 60 m² floor area. It may be required in smaller areas if there is additional hazard such as use by a large number of people who might obstruct signs.

The illumination of the open area shall be from two or more luminaires so that the failure of one luminaire does not result in total darkness or make the direction indication of the system ineffective.

The detailed requirements listed in clause 4.3 of BS EN 1838 are:

a The horizontal illuminance (density of luminosity on a flat plane) shall be not less than 0.5 lx at the floor level of the empty core area which excludes a border of 0.5 m of the perimeter of the area.

b The ratio of the maximum to the minimum anti-panic area lighting illuminance shall not be greater than 40:1.

c Disability glare shall be kept low by limiting the luminous intensity of the luminaires within the field of view. These shall not exceed the values in Table 1 of BS EN 1838 (Table 4.2 in this Guide) within the zone 60° to 90° from the downward vertical at all angles of azimuth (see Figure 2 of BS EN 1838).

d In order to identify safety colours, the minimum value for the colour rendering index R_a from a lamp shall be 40. The luminaire shall not substantially subtract from this.

e The minimum duration allowed for escape purposes shall be 1 h.

f The anti-panic area lighting shall reach 50% of the required illuminance within 5 s and full required illuminance within 60 s.

g Open area lighting is required to toilets for disabled people.

h Open area lighting is required in a room with no direct access to an escape route.

Spacing tables again are usually used to determine the number and position of additional luminaires, see Table 4.3.

▼ **Table 4.3**   Manufacturer's spacing table

Self-contained luminaires			Spacing, open area (m)			
Luminaire type	Mounting height (m)	Lux level directly under	0.5L ⊢─0	0──0.5L─0	0.5L	0.5L 0──⊣
Atlantic NM	2.5	1.66	2.5	9.1	5.4	1.4
NM	4.0	0.65	0.7	9.6	5.8	0.4
NM	6.0	0.29	–	–	–	–
M	6.0	0.29	2.4	8.8	5.0	0.4
M	4.0	0.55	0.9	8.0	5.0	0.6
M	6.0	0.25	–	–	–	–

[Courtesy of Menvier]

4.6 High risk task area lighting (clause 4.4 of BS EN 1838)

The objective of high risk task area lighting is to provide for the safety of people where there is a potentially dangerous process or situation such as to enable proper shut down procedures to be followed.

The detailed requirements of BS EN 1838 are:

a In areas of high risk the maintained illuminance on the reference plane shall be not less than 10% of the required maintained illuminance for that task, however it shall be not less than 15 lx. It shall be free of harmful stroboscopic effects.

b The uniformity* of the high risk task area lighting illuminance shall be not less than 0.1.

c Disability glare shall be kept low by limiting the luminous intensity of the luminaires within the field of view. These shall not exceed the values in Table 1 (Table 4.2 in this guide) within the zone 60° to 90° from the downward vertical at all angles of azimuth.

d In order to identify safety colours, the minimum value for the colour rendering index R_a of a lamp shall be 40. The luminaire shall not substantially subtract from this.

e The minimum duration shall be the period for which the risk exists to people.

f High risk task area lighting shall provide the full required illuminance permanently or within 0.5 s depending upon application.

The illumination required is normally achieved by installing a conversion of the luminaire to provide emergency lighting or by installing a tungsten projector unit.

The 10% required maintained illuminance for the task will normally be the limiting factor unless normal illumination levels are very low. Consequently, all fittings in the risk area need a standby ballast factor[†] of 10% or every other one of 20% etc.

4.7 Standby lighting (clause 4.5 of BS EN 1838)

Standby lighting is that part of emergency lighting provided to enable normal activities to continue. When it is used for emergency escape lighting purposes it must comply with the requirements for escape lighting.

Where a standby lighting level lower than the minimum normal lighting is employed, the lighting is to be used only to shut down or terminate processes.

---

* The uniformity of illumination is expressed in terms of the ratio of the minimum illuminance to the average illuminance.

† Ballast factor is a measure of the actual lumen output for a specific lamp-ballast system relative to the rated lumen output.

4

4.8 Classification of emergency lighting systems

(Annex C of BS 5266-1)

4.8.1 General

Emergency lighting systems are classified by:

▶ type of system
▶ mode of operation
▶ facilities, and
▶ duration of emergency mode.

Table 4.4 lists the detailed classifications under these four headings.

▼ **Table 4.4**   Emergency lighting classifications

Type	Mode of operation		Facilities		Duration of emergency mode for a self-contained system (minutes)
X self-contained	0	non-maintained	A	including test	10
Y central supply	1	maintained		device	60
	2	combined non-maintained	B	including remote rest mode	120
	3	combined maintained	C	including inhibiting mode	180
	4	compound non-maintained	D	high risk task area luminaire	
	5	compound maintained			
	6	satellite			

Until quite recently, emergency lighting systems were categorized by the prefix 'M' for maintained and 'NM' for non-maintained systems, followed by a '/' and the number of hours duration claimed for the installation, e.g. for self-contained systems:

M/1 was a maintained 1 h duration system; this is now

X	1	* * * *	60

NM/3 was a non-maintained 3 h duration system; this is now

X	0	* * * *	180

* * * * in the third box stands for the facilities, details of which are added, as applicable, at the time of installation.

4.8.2 Types of luminaire

X Self-contained emergency luminaire

Luminaire providing maintained or non-maintained emergency lighting in which all the elements, such as the battery, the lamp, the control unit and the test and monitoring facilities, where provided, are all contained within the luminaire or adjacent to it.

Y Centrally supplied emergency luminaire

Luminaire for maintained or non-maintained emergency lighting which is energized from a central emergency power system that is not contained within the luminaire.

4.8.3 Mode of operation

Non-maintained emergency luminaire

Luminaire in which the emergency lighting lamps are in operation only when the supply to the normal lighting fails.

Maintained emergency luminaire

Luminaire in which the emergency lighting lamps are energized at all times when normal or emergency lighting is required.

Combined emergency luminaire

Luminaire containing two or more lamps, at least one of which is energized from the emergency lighting supply and the other(s) from the normal lighting supply. A combined emergency luminaire is either maintained or non-maintained.

Sustained luminaire
This description is not found in BS 5266-1 or -8 or BS EN 1838, but is used to describe a combined emergency luminaire with two or more lamps where at least one lamp operates in non-maintained mode and is only illuminated when the normal supply fails. The other lamp operates on the normal supply only. This is identical in functionality to having a Non-Maintained Luminaire and a Normal Luminaire both in the same housing.

Compound self-contained emergency luminaire

Luminaire providing maintained or non-maintained emergency lighting and also providing emergency supply for operating a satellite luminaire.

4.8.4 Facilities

Remote rest mode

Feature of a self-contained emergency luminaire that can be intentionally extinguished by a remote device when the normal supply has failed and that, in the event of restoration of the normal supply, automatically reverts to normal mode.

4

Inhibiting mode

Feature of a self-contained emergency luminaire that can be set independently from the condition of the normal power and therefore, when the building is unoccupied, a supply failure will not cause unwanted discharge.

4.8.5 Duration

(clause 9 of BS 5266-1)

The classification (type, mode and duration) required for the system is agreed following the risk assessment and following consultation with the regulatory bodies. Guidance on minimum duration is given in BS 5266-1 and summarised in Table 4.5. The overriding consideration for the duration is that it is sufficient for the escape strategy.

▼ **Table 4.5** Emergency escape lighting duration

Class	Examples	Duration
Premises used as sleeping accommodation	This class includes such premises as hospitals, care homes, hotels, guest houses, certain clubs, colleges and boarding schools	3 h (note 1)
Non-residential premises used for treatment or care	This class includes such premises as special schools, clinics and similar premises	3 h
Non-residential premises used for recreation	This class includes such premises as theatres, cinemas, concert halls, exhibition halls, sports halls, public houses and restaurants	3 h (note 2)
Non-residential premises used for teaching, training and research, and offices	This class includes such premises as schools, colleges, technical institutes and laboratories	1 h
	If any part of a building is used outside of normal weekday office hours	3 h
Non-residential public premises	This class includes such premises as town halls, libraries, shops, shopping malls, art galleries and museums	3 h
Industrial premises used for manufacture, processing or storage of products	This class includes such premises as factories, workshops, warehouses and similar establishments	1 h
Multiple use of premises		As for the most stringent class
Common access routes within blocks of flats or maisonettes		3 h
Covered car parks		3 h
Sports stadia	for spectators	3 h

52 | Electrician's Guide to Emergency Lighting
© The Institution of Engineering and Technology

Notes to Table 4.5:

1 Guidance for hospitals is given in the Department of Health *Electrical services* series of publications (see [31], [32] and [33] in the Bibliography to BS 5266-1). Advice on the application of this guidance in relation to that given in this standard should be obtained from the enforcing authority.

2 Where the normal lighting might be dimmed or turned off, a maintained or combined emergency escape lighting system should be installed. However, it is not necessary for the full emergency lighting level to be provided when the normal lighting system is functioning. Full details of lighting requirements for places of entertainment are given in *Technical standards for places of entertainment* (see [36] in the Bibliography to BS 5266-1).

4.9 Requirements for safety signs

Minimum luminance

BS EN 1838 specifies requirements for:

▶ colour
▶ luminance
▶ ratios of colours
▶ ratios of luminance.

These are best achieved by using luminaires complying with BS EN 60598-2-22.

▼ **Figure 4.4**   Escape route sign (see also Figure 7.6)

The measurement of luminance and luminance ratios is not something that is easily done on site and should preferably be assured by the manufacturer.

Maximum viewing distances are:

▶ Internally illuminated signs 200 x height of pictogram (luminaire facia height)
▶ Externally illuminated signs 100 x height of pictogram (luminaire facia height)
▶ For example, an internally illuminated exit sign with pictogram height of 20 cm has a maximum viewing distance of 200 x 20 cm, that is 40 m. See Figure 4.5.

▼ **Figure 4.5** Escape route sign viewing distance

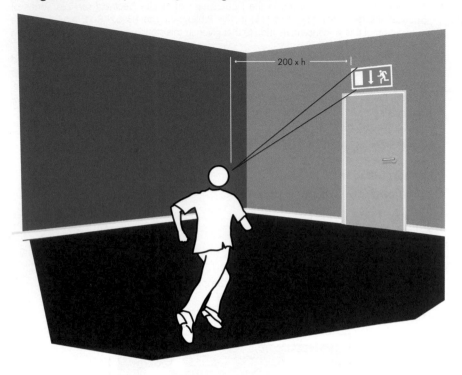

(200 x h)

4.10 Automatic test systems

4.10.1 Introduction

Automatic test systems are available for emergency lighting luminaires. Luminaires with a diagnostic interface unit can be specified. These can allow automatic testing on an individual luminaire basis or automatic testing as initiated by a central control unit.

Such test systems can be cost-effective as maintenance regimes are otherwise labour intensive.

(See also section 6.11 for standard types and user requirements.)

4.10.2 Central control testing systems

All the luminaires are fitted with addressable interface modules and connected by a communication medium such as data bus cable, see Figure 4.6. They can control both self-contained and central power systems.

▼ **Figure 4.6**   Automatic test system

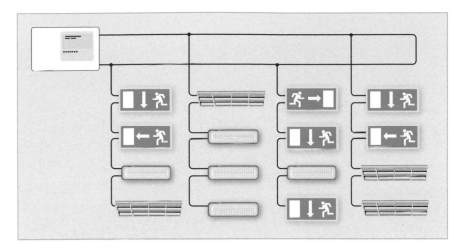

Note: The system may be fitted with data cable isolators, printer connections, PC interface etc.

Luminaires can be grouped for testing purposes so that, in the event of power failure after testing, a complete area is not left with run-down batteries, but the discharged luminaires are distributed around the building so maintaining effective signs and illumination.

See also section 6.11.

4.10.3 Stand-alone luminaire testing

Luminaires fitted with a suitable testing module can be programmed to carry out tests on an individual luminaire basis giving an external signal if a fault is detected, or have testing initiated by a hand-held device.

Luminaires can be individually programmed to carry out maintenance checks by a hand-held programmer. This type of system is suitable for use with self-contained luminaires only.

Electrical installation

5

5.1 Basic requirement

(clause 4.1 of BS 5266-8)

It is a requirement of BS 5266-8 that emergency escape lighting shall be activated not only on complete failure of the supply to the normal lighting but also on localized failure, such as a final circuit failure. Consequently, if the emergency escape lighting (that is non-maintained, self-contained or non-maintained from a central battery system) is supplied from a central supply there must be detectors in final lighting circuits to detect loss of supply.

5.2 Self-contained luminaires

(clause 8.1 of BS 5266-1)

Self-contained emergency luminaires do not require fire-protected cable supplies.

X type, or self-contained, luminaires have only one special requirement for their wiring other than that it must comply with BS 7671. Preferably, the wiring to the self-contained unit should not include or incorporate a plug and socket, unless precautions are taken to prevent accidental (or malicious) disconnection. There is no requirement to comply with the special requirements of BS 7671, chapter 56 (safety services) for the wiring to self-contained luminaires.

Often it is recommended that the wiring to self-contained emergency luminaires be the same as that for the normal lighting because if fire-resistant cabling is used for the emergency luminaires but not for the normal luminaires, a fire may cause failure of supply to the normal luminaires but not to the emergency luminaires and then in unusual circumstances they may not switch on

5

5.3 Central power supply systems (clause 8.2 of BS 5266-1)

5.3.1 General

The wiring of central supply installations must comply with BS 7671. Chapter 56, safety services, would apply to the wiring for central supply systems and there are, additionally, particular recommendations in BS 5266-1 (clause 8) for:

i   fire protection of the cables
ii   minimum cable size (1.5 mm^2)
iii voltage drop (4% maximum)
iv   conduit (metal or fire-resistant)
v   segregation
vi   joints
vii isolation and switching, and
viii labelling.

These recommendations are considered below.

Note: There may be a requirement for low emission of smoke and corrosive gases, as in section 20 of the London Building Acts as amended.

5.3.2 Wiring systems (clause 8.2.2 of BS 5266-1)

BS 5266-1 requires cables or cable systems with an inherently high resistance to attack by fire for connecting centrally supplied emergency escape lighting to the standby supply. Such cables and cable systems should have duration of survival of 60 minutes for standard systems and 120 minutes for enhanced systems when tested in accordance with BS EN 50200:2006.

This corresponds for standard emergency lighting systems to a classification of PH 60 as detailed in BS EN 50200:2006, and a duration of survival of 30 min when tested in accordance with BS EN 50200:2006, Annex E. For enhanced emergency lighting systems this corresponds to a classification of PH 120 as detailed in BS EN 50200: 2006 and a duration of survival of 120 min when tested in accordance with BS 8434-2.

Cables

BS 5266-1, clause 8.2.2 advises that cables with an inherently high resistance to attack by fire that may be suitable include:

1 Mineral insulated cables to BS EN 60702-1, with terminations complying with BS EN 60702-2
2 Fire-resistant electric cables having low emission of smoke and corrosive gases to BS 7629
3 Armoured fire-resistant electric cables having thermosetting insulation and low emission of smoke and gases to BS 7846.

It is necessary to seek confirmation from the cable supplier that these cables meet the requirement for either standard or enhanced systems, as no cable standard specifically calls up this test.

(It is understood that BS 7629-1 is currently under revision and will take into account standard and enhanced requirements.)

Note: These cables meet the recommendations of Approved Document B, see section 2.3.3 of chapter 2, provided they follow a route selected to pass only through parts of the building in which the fire risk is 'negligible' and they 'should' be separate from any circuit provided for another purpose.

Cable systems

BS 5266-1, clause 8.2.2c states that cable systems with an inherently high resistance to attack by fire that may be suitable include:

Fire-resistant cables enclosed in screwed steel conduit, such that the cable system has a duration of survival of 60 min when tested in accordance with IEC 60331-3.

Cable installers must confirm with their supplier that the cables do meet the duration of survival time of 60 minutes.

BS 5266-1, clause 8.2.2 notes that emergency escape lighting systems for certain large and complex buildings might require enhanced emergency lighting cables or cable systems capable of longer fire survival times, which might include unsprinklered buildings or parts of buildings.

Additional fire protection may be provided by the building structure if the cables are buried in the structure or installed where there is negligible fire risk and separated from any significant fire risk by a wall, partition or floor having at least a one hour fire resistance.

5.3.3 Cable cross-sectional area and voltage drop

(clauses 8.2.7 and 8.3.5 of BS 5266-1)

Cable conductors are required to be of copper with a minimum nominal cross-sectional area of 1.5 mm^2.

Voltage drop between the central battery or generator and slave luminaires should not exceed 4 per cent of the system nominal voltage at the maximum rated current and at the highest working temperature.

5.4 Cable support, fixings and joints

5.4.1 Cable support and fixings

(clause 8.2.3 of BS 5266-1)

Methods of cable support and fixings should be non-combustible, such that circuit integrity is not reduced below that afforded by the cable used. That is, they should be able to withstand a similar temperature and duration as the cable with the same water testing.

Note: In effect, this recommendation precludes the use of plastic supports and fixings where these would be the sole means of cable support.

Where support is provided by drop rods, the drop rod size should be calculated in accordance with BS 8519: *Code of practice for selection of fire resistant power and control cable systems for life-safety and fire-fighting applications.*

5.4.2 Cable joints

(clause 8.2.4 of BS 5266-1)

Cables should be installed without joints external to the equipment. Joints other than those within the system components such as the luminaires or control units should be constructed so that they have the same fire withstand (water and temperature) as the cable. Such a joint will need to be enclosed in a suitable box and labelled appropriately 'EMERGENCY LIGHTING', 'EMERGENCY ESCAPE LIGHTING' or 'STANDBY LIGHTING' and also with the warning 'MAY BE LIVE'. (See also regulation 537.2.1.3 of BS 7671 requiring a warning notice for any equipment supplied from more than one source. Emergency lighting circuits may be live when other circuits have been isolated.)

5.5 Segregation

(clause 8.2.6 of BS 5266-1)

For central supply systems it is essential that the wiring to the emergency escape lighting is segregated from the wiring of other circuits, either by installation in a separate conduit, ducting or trunking or by a metal or other rigid, mechanically strong and continuous partition of full depth without perforation.

5.6 Continuity of supply to the emergency lighting

(clause 8.3.1 of BS 5266-1)

The installation designer must bear in mind that maintenance and repair of the electrical installation in general will require from time to time circuits of the electrical installation to be switched off. Switching off the supply to emergency lighting luminaires may switch on the standby supply and run down the batteries, and may as a consequence in the event of a fire put the occupants at risk. The installation needs to be arranged so that, as far as is practicable, maintenance to the electrical installation can be carried out without disconnecting the supply to the emergency lighting. This will generally mean that for an installation comprising self-contained units the lighting circuits will need to be supplied separately from power circuits – see Figure 5.1. The occupier may also wish to switch off the building supply for safety or energy saving. However, such switching must not switch off the supply to the emergency lighting (so as to run down the standby batteries).

▼ **Figure 5.1** Circuit arrangement for self-contained luminaires

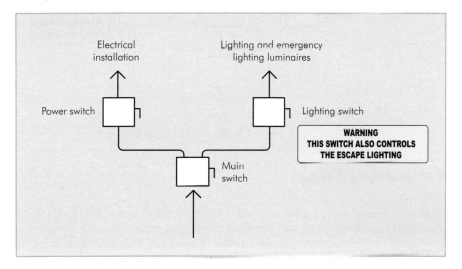

5.7 Isolation

(clause 8.3.2 of BS 5266-1)

For central supply systems, each isolating and switching device, etc. should be marked 'EMERGENCY LIGHTING', 'EMERGENCY ESCAPE LIGHTING' or 'STANDBY LIGHTING', as appropriate. For such systems it should be practicable to allow switching off of the general installation whilst maintaining the supply to the emergency lighting, in a similar way as it is necessary to maintain the supply to fire detection and alarm systems – see Figure 5.2.

▼ **Figure 5.2**   Circuit arrangement for central supply emergency lighting

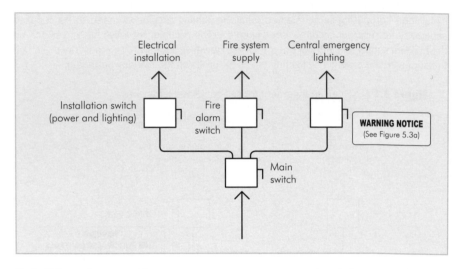

5.8 Warning notices

(regulation 514.11 of BS 7671 and clause 8.3.4 of BS 5266-1)

There is a general hazard associated with standby supply systems that switching off the circuits may not result in dead circuits as the standby supply may well make circuits live. Suitable warning notices must be installed. It must also be remembered that non-illumination of a luminaire or sign does not always indicate that the sign or circuit is dead. See Figure 5.3.

▼ **Figure 5.3**   Warning notices

a

```
WARNING
This equipment is also supplied from a standby supply.
Switching off will not make all circuits dead.
```

b

```
WARNING
Non-maintained luminaire.
Live when lamp not illuminated.
```

5.9 Inspection and testing

Installations must be inspected and tested in accordance with the requirements of BS 7671:2008 (as amended) *Requirements for Electrical Installations*, and an installation certificate issued together with inspection and test schedules.

Inspection and testing as per BS 5266-1 (clause 12) is necessary, see chapter 6.

5.10   BS 7671:2008 as amended

Chapter 56 of BS 7671 has requirements for 'safety services'. These are generally as per BS 5266-1 where applicable to emergency lighting.

Operation and maintenance   6

6.1 Disabled persons

A fire risk safety assessment guide (*Means of escape for disabled people*) has been published by the government to assist in the safe evacuation of disabled persons from non-residential premises. It advises on the preparation and practice of plans for evacuation of disabled people, including staff, by the use of Personal Emergency Evacuation Plans and plans for visitors.

6.2 Legislation   (regulation 4(2) of EWR)

The Electricity at Work Regulations 1989 (EWR) impose duties upon the employer, including:

> *As may be necessary to prevent danger, all systems shall be maintained so as to prevent, so far as is reasonably practicable, such danger*

The HSE *Memorandum of guidance on the Electricity at Work Regulations* requires regular inspection and maintenance and records of maintenance including test results.

6.3 Instructions   (clause 10.6 of BS 5266-1)

Unusually, British Standard 5266-1 recommends that the designer prepare instructions on the operation and maintenance of the emergency lighting. It is intended that the instructions should be in the form of a manual to be retained by the occupier of the building.

This manual is essential and must be obtained to form the starting point for the preparation of operation and maintenance procedures.

It is particularly important that operating instructions and manuals are provided for automatic testing systems.

6.4 As-installed drawings   (clause 6.1 of BS 5266-8)

On completion of the work, 'as-installed' drawings of the emergency escape lighting installation are required to be prepared for retention on the premises. The drawings shall identify all luminaires and the main components of the installation. The drawings shall

© The Institution of Engineering and Technology

be signed by a competent person to verify that they are 'as installed', and that the design meets the requirements of BS 5266-8.

Maintenance of the installation in accordance with BS 5266-8 requires that the drawings be updated as and when any changes or modifications are made to the installation. The drawings are required to be signed by a competent person to verify that the design meets the requirements of the standard.

The drawings are required to meet the requirements of regulation 514.9.1 of BS 7671. It is to be noted that for simple installations, this regulation requires a durable copy of a schedule relating to each particular distribution board to be provided within or adjacent to each distribution board.

6.5 Handover (clause 10.7 of BS 5266-1)

A competent person representing the designer and installer of the system on handover should:

i train a person or persons representing the client on the necessary routine inspections and tests or as a minimum how to use the automatic test features
ii provide necessary documentation for proper use of the system including design, installation and verification declarations, certification, logbook (see clause 11) and any other relevant information
iii explain the importance of keeping the logbook up to date and retaining the documents in an accessible safe place for reference as necessary
iv provide contact details.

6.6 Logbook (clauses 6.1 and 6.3 of BS 5266-8)

The standard requires a logbook to be retained on the premises in the care of a responsible person appointed by the occupier or owner. (The *Memorandum of guidance on the Electricity at Work Regulations* imposes this responsibility on the dutyholder.) The standard requires that the logbook will be used to record at least the following information:

a commissioning of the system
b each periodic inspection and test
c each service, inspection or test
d defects reported and date rectified
e any alteration (including additions) to the emergency lighting installation
f if any automatic testing device is employed, the main characteristic and the mode of operation of that device shall be described.

The user instructions and drawings prepared by the installer and approved by the designer should be retained with the logbook. If any automatic testing device is installed necessary information on its operation must be included in the instructions.

The logbook is intended to be used for all safety record-keeping, including that of the fire detection and alarm system, in particular all servicing, defects reported, repairs, operation of detectors and alarms including false alarms, attendance of fire brigade etc.

It should also record details of replacement of components, such as luminaires or batteries.

6.7 Care of batteries during installation

Repeated disconnection of emergency lighting during electrical installation can damage lamps and batteries. So far as is practicable, connection to the emergency lighting should be made only when supplies are unlikely to be regularly disconnected.

6.8 Servicing

6.8.1 Supervision (clause 7.1 of BS 5266-8)

BS 5266-8 requires the owner/occupier of the premises to appoint a competent person to supervise servicing. The Regulatory Reform (Fire Safety) Order 2005 puts responsibilities on the responsible person (usually the employer) for maintenance of the fire system, including escape lighting. See regulations 17 and 3 of the Order below.

17. Maintenance

(1) Where necessary in order to safeguard the safety of relevant persons the responsible person must ensure that the premises and any facilities, equipment and devices provided in respect of the premises under this Order or, subject to paragraph (6), under any other enactment, including any enactment repealed or revoked by this Order, are subject to a suitable system of maintenance and are maintained in an efficient state, in efficient working order and in good repair.

3. Meaning of 'responsible person'

In this Order 'responsible person' means –

(a) In relation to a workplace, the employer, if the workplace is to any extent under his control;

(b) in relation to any premises not falling within paragraph (a) –

(i) the person who has control of the premises (as occupier or otherwise) in connection with the carrying on by him of a trade, business or other undertaking (for profit or not); or

(ii) the owner, where the person in control of the premises does not have control in connection with the carrying on by that person of a trade, business or other undertaking.

6.8.2 Central batteries

(clause 12.1 of BS 5266-1)

The manufacturer's instructions are to be followed. Maintenance includes:

a keeping batteries and their terminals clean
b checking for leaks
c topping up the electrolyte.

Care must be taken that any replacement battery is compatible with the battery charger so that charging rates and times are appropriate; similarly, replacement chargers must be compatible with the batteries. Motor vehicle batteries are unlikely to be suitable.

Generally central batteries can last 15 to 20 years, while self-contained luminaire batteries will need changing every 5 years.

6.8.3 Generators

(clause 12.2 of BS 5266-1)

The manufacturer's instructions must be obtained, retained and followed. Poor maintenance will often result in failure of the emergency lighting.

Generators and control systems require regular maintenance, including test runs on load.

6.9 Routine inspection and testing

(clause 12.3 of BS 5266-1
and 7.2 of BS 5266-8)

Emergency lighting systems should be inspected and tested at regular intervals. The testing may be carried out manually or if this is not practicable, by using an automatic system.

BS 5266-1 notes that advice on conducting routine tests will have been given to the user as part of the handover procedure.

Note: When automatic testing is incorporated in the system, the test results shall be recorded monthly. For manual systems the frequency shall be as described below.

6.9.1 Daily

(clause 7.2.2 of BS 5266-8)

Visual inspection of central control indicators to check that the system is in an operational condition and requires no testing (this may be a duty of security patrols).

6.9.2 Monthly

(clause 7.2.3 of BS 5266-8)

If automatic testing is incorporated, the results of these automated tests must be recorded. For manual systems the following checks shall be carried out.

a By simulation of failure of the supply to the normal lighting, switch on in the emergency mode each luminaire and each internally illuminated sign for a period sufficient only to check that each lamp is illuminated. (Note: The period of simulated failure needs to be sufficient to check that all the luminaires are illuminated but not such as to impose a drain on the batteries.)
b At the end of this test period, the supply to the normal lighting is restored and every indicator lamp or device checked to ensure that it is showing that the normal supply has been restored.

c For central supply systems, correct operation of the system monitors is checked. For standby supplies to central systems such as batteries and generators, checks are made on the standby supply as recommended by the supplier. Any failure of the generators to start must be rectified immediately.

d The tests and results are recorded in the logbook.

6.9.3 Annually (clause 7.2.4 of BS 5266-8)

Where automatic testing facilities are installed the results of the full rated duration tests shall be recorded. For manual systems the following checks shall be carried out.

a By simulation of failure of the supply to the normal lighting, switch on in the emergency mode each luminaire and each internally illuminated sign for its full rated duration.

b At the end of this test period, the supply to the normal lighting is restored and every indicator lamp or device checked to ensure that it is showing that the normal supply has been restored. The charging arrangements are checked insofar as it is possible to do so.

c For central supply systems, correct operation of the system monitors is checked.

d For standby supplies to central systems such as batteries and generators, checks are made on the standby supply as recommended by the supplier.

e For generating sets the requirements of ISO 8528-12 are to be met.

f The tests and results are recorded in the logbook.

g A certificate is supplied to the person ordering the work and/or responsible for the safety systems of the premises.

Note: The fire risk assessment required by regulation 9 of the Regulatory Reform (Fire Safety) Order may result in a decision to increase maintenance frequencies.

6.10 Certificates

6.10.1 Completion certificates (regulation 632.1 of BS 7671 and clause 11 of BS 5266-1)

On completion of the installation of an emergency lighting system or an alteration or amendment, completion certificates together with schedules should be supplied to the person ordering the work, for forwarding to the occupier or owner of the premises. A BS 7671 certificate and schedules are required for the wiring and a separate certificate for the emergency lighting (see Annex A to this chapter).

On receiving the certificates it should be checked that all the information required by the British Standard model forms is included.

6.10.2 Periodic inspection and test certificate (clause 6.2 of BS 5266-8)

On completion of the yearly inspection and test a periodic inspection and test certificate shall be provided to the person ordering the work for retention on the premises. Many organizations provide preprinted forms and electronic forms. All are probably suitable but need to be checked against the model forms from the standard. See Annex C to this chapter.

6.11   Automatic testing systems (ATS)

BS EN 62034 *Automatic test system for battery-powered emergency escape lighting*

6.11.1 Classification of ATS types

The BS EN standard classifies automatic testing systems (ATS) as follows.

Type S   This is a stand-alone ATS consisting of a self-contained luminaire with a built-in testing facility that provides a local indication of the condition of the luminaire, but still requires all luminaires to be manually inspected, with a manual record made of the information indicated by the luminaires.

Type P   The emergency luminaires are monitored and their condition is indicated by a test facility that collects and displays the results of the tests, but requires manual recording of information on the tests.

Type ER   As Type P, but the test facility collects results, and data is recorded and logged by the ATS.

Type PER   As types P or ER, but with a collated fault indicator that automatically gives remote indication of failure of any of the luminaires that have been tested.

TYPE PERC As type PER, but with the additional features of a central controller, for setting parameters, configuration of the system and the central controlled initialization of the test and where the date, time and duration of the test is defined by the central controller.

The introduction to the standard advises that automatic test systems will still require manual intervention to correct faults when they are identified, and procedures should be put in place for such intervention.

A visual check of system components and indicators should be included in the routine of safety staff. This check should be made regularly to ensure that emergency luminaires remain present and intact.

6.12   The Waste Electrical and Electronic Equipment (WEEE) Directive

The Waste Electrical and Electronic Equipment (WEEE) Directive came into force in January 2007 and aims to both reduce the amount of WEEE being produced and encourage everyone to reuse, recycle and recover it. The WEEE Directive also aims to improve the environmental performance of businesses that manufacture, supply, use, recycle or recover electrical and electronic equipment.

WEEE legislation introduced new responsibilities for businesses and other non-household users of electrical and electronic equipment (EEE). This includes businesses, schools, hospitals and government agencies, when they dispose of their electrical waste. These organisations need to ensure that all separately collected WEEE is treated and recycled. Whether the business or the producer of the EEE pays for this, depends on the circumstances. Gas discharge lamps and LEDs and the luminaires are within the scope of the WEEE Directive.

Factsheets are available from the Environment Agency website.

6.13 Guidance

The guidance given in this chapter is not exhaustive. Further guidance is available in the publications of BSRIA, CIBSE and HSE, for example:

▶ BSRIA – *Illustrated guide to electrical building services BG5/2005*
▶ BSRIA – *Handover, O&M Manuals and Project Feedback: A toolkit for designers and contractors BG1/2007*
▶ CIBSE – Lighting Design Guide 12: *Emergency lighting design guide – Guide to ownership, operation and maintenance of building services*

Annex A – Model completion certificate

A1   Emergency lighting completion certificate

Serial Number:...................

EMERGENCY LIGHTING COMPLETION CERTIFICATE

For New Installations

Occupier/owner...

Address of premises ....

....

Declaration of Conformity

In consequence of acceptance of the appended declarations, I/we* hereby declare that the emergency lighting system installation, or part thereof, at the above premises conforms, to the best of my/our* knowledge and belief, to the appropriate recommendations given in BS 5266-1:2011, *Emergency lighting – Part 1: Code of practice for the emergency lighting of premises*, BS EN 1838:1999 *Lighting applications – Emergency lighting* and BS EN 50172:2004, *Emergency escape lighting systems*, as set out in the accompanying declarations, except as stated below/overleaf.

* Delete as appropriate.

Signed, on behalf of owner/occupier ....

Name...

Deviations from standards

Declaration (Design, installation or verification)	Clause number	Details of deviation

This Certificate is only valid when accompanied by current:

a)   Signed declaration(s) of design, installation and verification, as applicable (see overleaf).

b)   Photometric design data. This can be in any of the following formats but in all cases appropriate de-rating factors must be used and identified to meet worst case requirements.
   • Authenticated spacing data such as ICEL 1001 registered tables**.
   • Calculations as detailed in Annex E and CIBSE/SLL Guide LG12***.
   • Appropriate computer print of results.

c)   Test log book.

**Available from Industry Committee for Emergency Lighting, Ground Floor, Westminster Tower, 3 Albert Embankment, London, SE1 7SL.

***Available from Chartered Institution of Building Services Engineers, Delta House, 222 Balham High Road, London SW12 9BS.

A2   Design: declaration of conformity

BS 5266-1: 2011 clause reference	Recommendations	System conforms (if NO, record a deviation)		
		YES	NO	N/A
4.2	**D1** Accurate plans available showing escape routes, fire alarm control panel, call points and fire extinguishers			
5.4	**D2** Fire safety signs in accordance with BS 5499-4, and other safety signs in accordance with BS ISO 7010, clearly visible and adequately illuminated			
6.7	**D3** The luminaires conform to BS EN 60598-2-22			
6.6	**D4** Luminaires located at following positions: NOTE Near means within 2 m horizontally. a) At each exit door intended to be used in an emergency b) Near stairs so each tread receives direct light, and near any other level change c) Near mandatory emergency exits and safety signs d) At each change of direction and at intersections of corridors e) Outside and near to each final exit f) Near each first aid post g) Near fire-fighting equipment and call points			
6.3	**D5** At least two luminaires illuminating each compartment of the escape route			
	D6 Additional emergency lighting provided where needed to illuminate:			
6.6.3	a) Lift cars			
6.6.4	b) Moving stairways and walkways			
6.6.5	c) Toilet facilities larger than 8 m² floor area or without borrowed light, and those for disabled use			
6.6.6	d) Motor generator, control and plant-rooms			
6.6.7	e) Covered car parks			
9.1	**D7** Design duration adequate for the application			
10.6; 10.7; Clause 11	**D8** Operation and maintenance instructions and a suitable log book produced for retention and use by the building occupier			
5.1.1; 5.1.2	**D9** Illuminance. Escape routes for any use: 1 lx min. on the centre line Open areas above 60 m²: 0.5 lx min. anywhere in the core area			

Serial Number:.................

Design – Declaration of conformity

Deviations from standards
(to be entered on Completion Certificate)

Clause number	Details of deviation

Signature of person making design conformity declaration...

For and on behalf of .... Date...

A3   Installation: declaration of conformity

Serial Number:.................

Installation – Declaration of conformity

BS 5266-1: 2011 clause reference	Recommendations	System conforms (if NO, record a deviation)		
		YES	NO	N/A
Clause 5	IN1 The system installed conforms to the agreed design			
6.1	IN2 All non-maintained luminaires fed or controlled by the final circuit supply of their local normal mains lighting			
6.4	IN3 Luminaires mounted at least 2 m above the floor			
6.4	IN4 Luminaires mounted at a suitable height to avoid being located in smoke reservoirs or other likely area of smoke accumulation			
5.4	IN5 Fire safety signs in accordance with BS 5499-4, and other safety signs in accordance with BS ISO 7010, clearly visible and adequately illuminated			
8.2	IN6 The wiring of central power systems has adequate fire protection and is appropriately sized			
8.3.5	IN7 Output voltage range of the central power system is compatible with the supply voltage range of the luminaries, taking into account supply cable voltage drop			
8.2.12	IN8 All plugs and sockets protected against unauthorized use			
8.3.3	IN9 The system has suitable and appropriate testing facilities for the specific site			
Clause 11	IN10 The equipment manufacturers' installation and verification procedures satisfactorily completed			
Clause 8	IN11 The system conforms to BS 7671			

Deviations from standards
(to be entered on Completion Certificate)

Clause number	Details of deviation

Signature of person making design conformity declaration...

For and on behalf of .... Date...

A4   Verification: declaration of conformity

Serial Number:.................

Verification – Declaration of conformity

BS 5266-1: 2011 clause reference	Recommendations	System conforms (if NO, record a deviation)		
		YES	NO	N/A
4.2	**V1** Plans available and correct			
8.3.3	**V2** System has a suitable test facility for the application			
5.4	**V3** All escape route safety signs and fire-fighting equipment location signs, and other safety signs identified from risk assessment, visible with the normal lighting extinguished			
Clause 5	**V4** Luminaires correctly positioned and oriented as shown on the plans			
6.7 and Annex C	**V5** Luminaires conform to BS EN 60598-2-22			
6.7 and Annex C	**V6** Luminaires have an appropriate category of protection against ingress of moisture or foreign bodies for their location as specified in the system design			
12.3	**V7** Luminaires tested and found to operate for their full rated duration			
12.3	**V8** Under test conditions, adequate illumination provided for safe movement on the escape route and the open areas NOTE This can be checked by visual inspection and checking that the illumination from the luminaires is not obscured and that minimum design spacings have been met.			
12.3	**V9** After test, the charging indicators operate correctly			
8.2	**V10** Fire protection of central wiring systems satisfactory			
8.2.6	**V11** Emergency circuits correctly segregated from other supplies			
10.6; 10.7; Clause 11	**V12** Operation and maintenance instructions together with a suitable log book showing a satisfactory verification test provided for retention and use by the building occupier			
Additional recommendations for verification of an existing installation				
10.7 and Clause 12	**V13** Building occupier and their staff trained on suitable maintenance, testing and operating procedures, or a suitable maintenance contract agreed			
Clause 11	**V14** Test records in the log book complete and satisfactory			
Clause 12	**V15** Luminaires clean and undamaged with lamps in good condition			
Clause 6	**V16** Original design still valid NOTE If the original design is not available this needs to be recorded as a deviation.			

Deviations from standards
(to be entered on Completion Certificate)

Clause number	Details of deviation

Signature of person making design conformity declaration ....

For and on behalf of .... Date...

Annex B – Model certificate for completion of small new installations and verification of existing installations

B1 General declaration

Serial Number:..................

EMERGENCY LIGHTING SMALL* NEW INSTALLATIONS AND EXISTING SITE COMPLIANCE CERTIFICATE

For Small New Installations and Verification of Existing Installations

Occupier/owner..

Address of premises ....

....

Declaration of Conformity

In consequence of acceptance of the appended declarations, I/we* hereby declare that the emergency lighting system installation, or part thereof, at the above premises conforms, to the best of my/our* knowledge and belief, to the appropriate recommendations given in BS 5266-1:2011, *Emergency lighting – Part 1: Code of practice for the emergency lighting of premises*, BS EN 1838:1999 *Lighting applications – Emergency lighting* and BS EN 50172:2004, *Emergency escape lighting systems*, as set out in the accompanying declarations, except as stated below/overleaf.

* Delete as appropriate.

Signed, on behalf of owner/occupier ....

Name...

Deviations from standards

Declaration (Design, installation or verification)	Clause number	Details of deviation

This Certificate is only valid when accompanied by current:

a) Signed declaration(s) of design, installation and verification, as applicable (see overleaf).

b) Photometric design data. This can be in any of the following formats but in all cases appropriate de-rating factors must be used and identified to meet worst case requirements.
   - Authenticated spacing data such as ICEL 1001 registered tables**.
   - Calculations as detailed in Annex E and CIBSE/SLL Guide LG12***.
   - Appropriate computer print of results.

c) Test log book.

*New works are deemed to be small when involving installations of up to 25 new emergency lighting luminaires

**Available from Industry Committee for Emergency Lighting, Ground Floor, Westminster Tower, 3 Albert Embankment, London, SE1 7SL.

***Available from Chartered Institution of Building Services Engineers, Delta House, 222 Balham High Road, London SW12 9BS.

B2 Model certificate for completion of small new installations and verification of existing installations – Checklist and report

Site Address			Responsible Person			
BS 5266-1: 2011	Engineer Function D-Designer, I-Installer, V-Verifier		Inspection Date			
clause ref.	**D,I,V**	**Check of categories and documentation**		**YES**	**NO**	**N/A**
4.2	D,V	Are plans of the system available and correct?				
Clause 9	D,V	Has the system been designed for the correct mode of operation category?				
Clause 9	D,V	Has the system been designed for the correct emergency duration period?				
Clause 11	D,V	Is a completion certificate available with photometric design data?				
Clause 11	D,I,V	Is a test log book available and are the entries up to date?				
		Check of design				
Clause 6	D,I,V	Are the correct areas of the premises covered to meet the risk assessment?				
Clause 6	D,I,V	Are all hazards identified by the risk assessment covered?				
Clause 5	D,I,V	Are there luminaires sited at the "points of emphasis"?				
Clause 5	D,I,V	Is the spacing between luminaires compliant with authenticated spacing or design data?				
Clause 5	D,I,V	If authenticated spacing data is not available for existing installations, are estimates attached and acceptable?				
5.4	D,I,V	Are the emergency exit signs and arrow directions correct and the locations of other safety signs to be illuminated under emergency conditions identified?				
6.1	D,I,V	Do all non-maintained luminaires operate on local final circuit failure?				
6.3	D,V	Is there illumination from at least two luminaires in each compartment?				
6.4	I,V	Are luminaires at least 2 m above floor and avoiding smoke reservoirs?				
6.6	D,V	Are additional luminaires located to cover toilets, lifts, plant rooms etc?				
		Check of the quality of the system components and installation				
6.7	D,I,V	Do the luminaires conform to BS EN 60598-2-22?				
6.7	D,I,V	Do any converted luminaires conform to BS EN 60598-2-22?				
6.7	D,I,V	Do luminaires have a suitable degree of protection for their location?				
Clause 8	I,V	Does the installation conform to the good practice defined in BS 7671?				
8.2.1	D,I,V	For centrally powered systems, is the wiring fire-resistant?				
8.2.12	D,I,V	Are any plugs or sockets protected against unauthorized use?				
8.2.1	D,I,V	If a central power supply unit is used, does it conform to BS EN 50171?				
8.2.1	D,V	Can AC systems start the lamps from the battery in an emergency?				
8.2.1	D,V	Can AC systems clear all distribution fuses/miniature circuit breakers in an emergency?				

B2 Model certificate for completion of small new installations and verification of existing installations – Checklist and report
continued

		Test facilities	YES	NO	N/A
8.3.3	D,V,I	Are the test facilities suitable to test function and duration?			
8.3.3	D,I,V	Are the test facilities safe to operate and do not isolate a required service?			
8.3.3	D,I,V	Are the test facilities clearly marked with their function?			
8.3.3	D,I,V	If an automatic test system is installed, does it conform to IEC 62034?			
10.7	D,V	Are the user's staff trained and able to operate the test facilities and record the test results correctly?			
		Final acceptance to be conducted at completion			
12.3	D,I,V	Does the system operate correctly when tested?			
10.7	D,I,V	Has adequate documentation been provided to the user?			
10.7	D,I,V	Is the user aware of action they should take in the event of a test failure?			
10.7	D,I,V	Are any deviations fully documented and are they still acceptable?			

Action recommended or deviation to be reported:

Results of the inspection ....……   Signed..

....…….   ....

Annex C – Model periodic inspection and test certificate

C1   Model emergency lighting inspection and test certificate

Emergency Lighting Inspection and Test Certificate

For systems designed to BS 5266-1 and BS EN 50172/BS 5266-8

WARNING

Full duration tests involve discharging the batteries, so the emergency lighting system will not be fully functional until the batteries have had time to recharge. For this reason, always carry out testing at times of minimal risk, or only test alternate luminaires at any one time.

System manufacturer			
Contact phone number			
System installer			
Contact phone number			
Competent engineer responsible for verification and annual tests		Phone number	
Site address			
Responsible person			
Date the system was commissioned			
Details of system mode of operation	Non-maintained		
	Non-maintained luminaires, maintained signs		
	Maintained		
	Other		
Duration of system	.... Hours	Is automatic test system fitted?	Y/N

Details of additions or modifications to the system or the premises since original installation	
Addition or modification	**Date**

Action to be taken on finding a failure

- The supplier of the system or a competent engineer should be contacted to rectify the fault.

- A risk assessment of the failure should be conducted; this should evaluate the people who will be at increased risk and the level of that risk. Based on this data and, if necessary, advice from the Fire Authority, the appropriate action should be taken.

- Action may be:

  To warn occupants to be extra vigilant until the system is rectified

  To initiate extra safety patrols

  To issue torches as a temporary measure

  In a high risk situation, to limit use of all or part of the building

NOTE   Test programs for identifying early failures can reduce the chances of failure of two adjacent luminaires at the same time.

C 2   Inspection and test record

Emergency Lighting Inspection and Test Record			Sheet number:	
Site:				
Test types:	C = Commissioning and verification test			
	M = Monthly test (see BS EN 50172:2004/BS 5266-8:2004, **7.2.3**)			
	A = Annual test (see BS EN 50172:2004/BS 5266-8:2004, **7.2.4**)			
Date of test	**Test type**	**Result – Test Passed No action needed**	**Result – Test Failed**	
			Need for repair of system notified	**Need for safeguarding of premises notified**
		Sign below *	Sign below*	Sign below*
	C			
	M – 1st month			
	M – 2nd month			
	M – 3rd month			
	M – 4th month			
	M – 5th month			
	M – 6th month			
	M – 7th month			
	M – 8th month			
	M – 9th month			
	M – 10th month			
	M – 11th month			
	A – 1st year			
	M – 1st month			
	M – 2nd month			
	M – 3rd month			
	M – 4th month			
	M – 5th month			
	M – 6th month			
	M – 7th month			
	M – 8th month			
	M – 9th month			
	M – 10th month			
	M – 11th month			
	A – 2nd year			
	M – 1st month			
	M – 2nd month			
	M – 3rd month			
	M – 4th month			
	M – 5th month			
	M – 6th month			
	M – 7th month			
	M – 8th month			
	M – 9th month			
	M – 10th month			
	M – 11th month			
	A – 3rd year			

C3   Fault action record

Emergency Lighting Fault Action Record		Sheet number:		
Contact references	Contact name	Phone number		
Equipment supplier:			For replacement parts	
Maintenance engineer:			Competent engineer	
Responsible person:			Site control	
Date of failure	Action taken to safeguard the premises (Details and signature)	Action taken to rectify the system (Details and signature)		Date system repaired

Safety signs   7

7.1 Introduction

7.1.1 The Health and Safety (Safety Signs and Signals) Regulations 1996 (SI 1996 No. 341)

Where the risk assessment required by the Management of Health and Safety at Work Regulations indicates a need for any safety signs or signals, the Health and Safety (Safety Signs and Signals) Regulations require that suitable signs be installed including if necessary standby supplies.

Relevant parts of the Safety Signs and Signals Regulations are reproduced in section 1.5 of chapter 1.

The Health and Safety Executive have published guidance on the safety signs and signals of the Health and Safety (Safety Signs and Signals) Regulations (in publication L64).

In this chapter the safety signs associated with fire systems and escape lighting recommended in the publication are reproduced.

7.1.2 Approved Document B   (para 5.37 of Vol. 2)

The Approved Document recommends that except in a flat (as opposed to common areas), every escape route other than those in ordinary use should be distinctively and conspicuously marked by emergency exit signs of adequate size, complying with the Health and Safety (Safety Signs and Signals) Regulations 1996. Reference is also made to BS 5499-1 (but see clause 2.3.2 of this Guide).

7.2 Format of safety signs

Schedule 1 to the Health and Safety (Safety Signs and Signals) Regulations 1996 lays down the format of safety signs, including the shape and colour.

The shapes, etc. recommended in the guidance are as follows.

▼ **Figure 7.1**   Prohibitory signs: round shape, black pictogram on white background, red edging and diagonal line

▼ **Figure 7.2**   Warning signs: triangular shape, black pictogram on a yellow background with black edging

Figure 7.3 Mandatory signs: round shape, white pictogram on blue background

Figure 7.4 Emergency escape and first-aid signs: rectangular or square shape, white pictogram on a green background

Figure 7.5 Fire-fighting signs: rectangular or square shape, white pictogram on a red background

7.3 Emergency escape and first-aid signs

The signs given in HSE publication L64 *Safety signs and signals* are reproduced below.

▼ **Figure 7.6**   Emergency escape

▼ **Figure 7.7** Supplementary 'This way'

▼ **Figure 7.8** First aid

a First-aid post **b** Shower **c** Stretcher

d Eyewash **e** Emergency telephone

7

7.4 Fire-fighting signs

▼ **Figure 7.9** Fire-fighting signs

a Fire hose

b Ladder

c Emergency fire telephone

d Fire extinguisher

▼ **Figure 7.10** Supplementary 'This way' signs for fire-fighting equipment

Index

IET Wiring Regulations and associated publications

The IET prepares regulations for the safety of electrical installations, the IET Wiring Regulations (BS 7671 *Requirements for Electrical Installations*), which are the standard for the UK and many other countries. The IET also offers guidance around BS 7671 in the form of the Guidance Notes series and the Electrician's Guides as well as running a technical helpline. The Wiring Regulations and guidance are now also available as e-books through Wiring Regulations Digital (see overleaf).

IET Members receive discounts across IET publications and e-book packages.

Requirements for Electrical Installations, IET Wiring Regulations 17th Edition (BS 7671:2008 incorporating Amendment No. 1:2011)
Order book PWR1701B Paperback 2011
ISBN: 978-1-84919-269-9   **£80**

On-Site Guide BS 7671:2008 (2011)
Order book PWGO171B Paperback 2011
ISBN: 978-1-84919-287-3   **£24**

IET Guidance Notes
The IET also publishes a series of Guidance Notes enlarging upon and amplifying the particular requirements of a part of the IET Wiring Regulations.

Guidance Note 1: Selection & Erection, 6th Edition
Order book PWG1171B Paperback 2012
ISBN: 978-1-84919-271-2   **£72**

Guidance Note 2: Isolation & Switching, 6th Edition
Order book PWG2171B Paperback 2012
ISBN: 978-1-84919-273-6   **£27**

Guidance Note 3: Inspection & Testing, 6th Edition
Order book PWG3171D Paperback 2012
ISBN: 978-1-84919-275-0   **£27**

Guidance Note 4: Protection Against Fire, 6th Edition
Order book PWG4171B Paperback 2012
ISBN: 978-1-84919-277-4   **£27**

Guidance Note 5: Protection Against Electric Shock, 6th Edition
Order book PWG5171B Paperback 2012
ISBN: 978-1-84919-279-8   **£27**

Guidance Note 6: Protection Against Overcurrent, 6th Edition
Order book PWG6171B Paperback 2012
ISBN: 978-1-84919-281-1   **£27**

Guidance Note 7: Special Locations, 4th Edition
Order book PWG7171B Paperback 2012
ISBN: 978-1-84919-283-5   **£27**

Guidance Note 8: Earthing & Bonding, 2nd Edition
Order book PWG8171B Paperback 2012
ISBN: 978-1-84919-285-9   **£27**

Electrician's Guides

Electrician's Guide to the Building Regulations, 3rd Edition
Order book PWGP171B Paperback 2013
ISBN: 978-1-84919-556-0   **£24**

Electrician's Guide to Fire Detection and Alarm Systems, 2nd Edition
Order book PWR15130 Paperback 2013
ISBN: 978-1-84919-763-2   **£25**

continues ▶

Other guidance

**Commentary on IEE Wiring Regulations
(17th Edition, BS 7671:2008)**
Order book PWR08640 Hardback 2009
ISBN: 978-0-86341-966-9   **£65**

Electrical Maintenance, 2nd Edition
Order book PWR05100 Paperback 2006
ISBN: 978-0-86341-563-0   **£40**

**Code of Practice for In-service Inspection
and Testing of Electrical Equipment,
4th Edition**
Order book: PWR02340 Paperback 2012
ISBN: 978-1-84919-626-0   **£55**

**Electrical Craft Principles, Volume 1,
5th Edition**
Order book PBNS0330 Paperback 2009
ISBN: 978-0-86341-932-4   **£25**

**Electrical Craft Principles, Volume 2,
5th Edition**
Order book PBNS0340 Paperback 2009
ISBN: 978-0-86341-933-1   **£25**

For more information and to buy the IET Wiring
Regulations and associated guidance, visit
www.theiet.org/electrical

Wiring Regulations Digital: Pick&Mix

The IET Wiring Regulations and associated guidance
are now available in e-book format. You can search
and link within and between books, make your own
notes and print pages from the books. E-books are
available individually*, or as part of a package as
outlined below.

Wiring Regulations Digital: Essentials
BS 7671:2008 (2011), the On-Site Guide,
    £98.70 (inc VAT)

Wiring Regulations Digital: Business
BS 7671:2008 (2011), the On-Site Guide,
Guidance Notes 1–8, the Electrician's Guide to the
Building Regulations and the Electrical Installation
Design Guide.   **£330 (inc VAT)**

Wiring Regulations Digital: Online
For larger companies looking for multi-user
access to the Wiring Regulations e-books, Wiring
Regulations Digital: Online is the perfect solution.
Find out more at **www.theiet.org/wrdo**

Find out more and order at
www.theiet.org/digital-regs

* BS 7671 must be purchased with the On-Site Guide – buy
together for £98.70 (inc VAT).

IET Standards

IET Standards works with industry-leading bodies
and experts to publish a range of codes of practice
and guidance materials for professional engineers,
including:

**Code of Practice for Electrical Safety
Management**
Order book PSES001P Paperback 2013
ISBN: 978-1-84919-669-7   **£130**

**Code of Practice for the Application of LED
Lighting Systems**
Order book PSLS001P Paperback 2014
ISBN: 978-1-84919-719-9   **£55**

**Code of Practice for Electric Vehicle
Charging Equipment Installation**
Order book PSEV001P Paperback 2012
ISBN: 978-1-84919-514-0   **£55**

See the full range of titles at
www.theiet.org/standards

IET The Institution of
Engineering and Technology

The IET Centres of Excellence programme recognises training providers who consistently achieve high standards of training delivery. Look for a Centre of Excellence in order to:

- Be certain that your course has been independently quality assured
- Get a course that underpins the expertise required of the IET Wiring Regulations and other standards and codes of practice
- Have confidence in your trainers who have been approved by the industry experts at the IET

Find out more at **www.theiet.org/excellence**.

Dedicated website

Everything that you need from the IET is now in one place. Ensure that you are up to date with BS 7671 and find guidance by visiting our dedicated website for the electrical industry.

Catch up with the latest forum discussions, download the Wiring Regulations forms (as listed in BS 7671) and order e-books through Wiring Regulations Digital.

The IET Wiring Regulations BS 7671:2008 (2011) and all associated guidance publications can be bought directly from the site. You can even pre-order titles, that have not yet been published, at discounted prices.

www.theiet.org/electrical

Membership

Passionate about your career? Become an IET Member and benefit from a range of benefits from the industry experts. As co-publishers of the IET Wiring Regulations, we can assist you in demonstrating your technical professional competence and support you with all your training and career development needs.

The Institution of Engineering and Technology is the professional home for life for engineers and technicians. With over 150,000 members in 127 countries, the IET is the largest professional body of engineers in Europe.

Joining the IET and having access to tailored products and services will become invaluable for your career and can be your first step towards professional qualifications.

You can take advantage of ...

- a 35% discount on BS 7671:2008 (2011) IET Wiring Regulations, associated guidance publications and Wiring Regulations Digital: Pick&Mix
- career support services to assist throughout your professional development
- dedicated training courses, seminars and events covering a wide range of subjects and skills
- an array of specialist online communities
- professional development events covering a wide range of topics from essential business skills to time management, health and safety, life skills and many more
- access to over 100 Local Networks around the world
- live IET.tv event footage
- instant online access to over 70,000 books, 3,000 periodicals and full-text collections of electronic articles through the Virtual Library, wherever you are in the world.

Join online today: **www.theiet.org/join** or contact our membership and customer service centre on +44 (0)1438 765678.